ACS SYMPOSIUM SERIES 245

Size Exclusion Chromatography

Methodology and Characterization of Polymers and Related Materials

Theodore Provder, EDITOR

Glidden Coatings and Resins

Based on a symposium sponsored by
the Division of Organic Coatings
and Plastics Chemistry
at the 185th Meeting
of the American Chemical Society,
Seattle, Washington,
March 20–25, 1983

American Chemical Society, Washington, D.C. 1984

SEP/AE
CHEM
_69857921_

Library of Congress Cataloging in Publication Data

Size exclusion chromatography.
(ACS symposium series, ISSN 0097-6156; 245)

"Based on a symposium sponsored by the Division of
Organic Coatings and Plastics Chemistry at the 185th
Meeting of the American Chemical Society, Seattle,
Washington, March 20-25, 1983."

Includes bibliographies and indexes.

1. Gel permeation chromatography—Congresses.
2. Polymers and polymerization—Analysis—
Congresses.

I. Provder, Theodore, 1939-   . II. American
Chemical Society. Division of Organic Coatings and
Plastics Chemistry. III. Series.

QD272.C444S6   1984   547.7'046   83-27515
ISBN 0-8412-0826-3

ACS Symposium Series

M. Joan Comstock, *Series Editor*

Advisory Board

FOREWORD

The ACS SYMPOSIUM SERIES was founded in 1974 to provide
a medium for publishing symposia quickly in book form. The
format of the Series parallels that of the continuing ADVANCES
IN CHEMISTRY SERIES except that in order to save time the
papers are not typeset but are reproduced as they are submitted by the authors in camera-ready form. Papers are reviewed under the supervision of the Editors with the assistance
of the Series Advisory Board and are selected to maintain the
integrity of the symposia; however, verbatim reproductions of
previously published papers are not accepted. Both reviews
and reports of research are acceptable since symposia may
embrace both types of presentation.

CONTENTS

v

PREFACE

THE FIELD OF SIZE EXCLUSION CHROMATOGRAPHY (SEC) continues to grow in scope and in depth. Since the last American Chemical Society symposium on this subject in 1979, about 300 papers have been published annually. The continuing interest in the field is a result of (1) improved column technology, (2) availability of improved and varied in-line detectors, and (3) improved data treatment procedures and methods facilitated by the microcomputer explosion of the last 5 years.

This volume deals with the methodology involved in the practice of SEC from both a theoretical and a pragmatic perspective along with the application of this methodology to the characterization of polymers and related materials. The three sections reflect the major efforts in the field over the last 3 years.

In the first section, the mechanisms involved in size exclusion chromatography are discussed; this is an area where additional understanding and clarification still are needed. Data treatment with respect to statistical reliability of the data along with corrections for instrumental broadening is still a valid concern. Instrumental advances in the automation of multiple detectors and the development of a pressure-programmed, controlled-flow supercritical fluid chromatograph are presented.

In the second section, improved column technology is emphasized. The effects of operational variables on the performance of the chromatographic system are considered. Some of the operational variable concerns are shear degradation of high molecular weight polymers, the use of mixed solvent systems, and the optimization of resolution for analysis of oligomers and small molecules.

In the third section, the emphasis is on the application of SEC methodology for the characterization of polymers. The use of continuous in-line low-angle laser light-scattering detection is illustrated for the high-temperature SEC analysis of polyethylene and of linear and branched block copolymers. The development of a continuous in-line viscosity detector and its application as an absolute molecular weight detector is described. The application of SEC for cross-linked network analysis by studying thermoset resin cure kinetics and cross-linked network morphology is of special interest.

This book has brought together papers that represent current activity in the field of SEC. It is hoped that this book will spur further activity in the field.

Special thanks are given to the authors for their effective oral and written communication and to the reviewers for their critiques and constructive comments.

THEODORE PROVDER
Glidden Coatings and Resins
Strongsville, Ohio

December 1983

MECHANISM, INSTRUMENTATION, CALIBRATION, AND DATA TREATMENT

Mathematical Modeling of Particle Chromatography

D. C. FRANCIS and A. J. McHUGH

Department of Chemical Engineering, University of Illinois, Urbana, IL 61801

A discussion is given of the mathematical modelling of the separation mechanisms associated with the packed column chromatography of particulate systems. Primary emphasis is on the derivation of models for the HDC and pore partitioning processes which occur with porous packing systems. Comparison is made of predictions for the separation factor - particle size behavior for a purely flow-through model, published earlier, and models developed herein to account for simultaneous pore partitioning effects. Comparison to literature data indicates that accounting for pore partitioning leads to a more accurate fit. The results of these calculations indicate the need for further experimental studies to characterize the model parameters associated with the possible separation mechanisms.

A large and important class of colloids are the polymer latexes which consist of charged (by ionogenic surface groups and/or adsorbed species) generally spherical particles with diameters ranging from tens of nanometers to microns. The role of particle size analysis in characterizing such systems, for both fundamental studies and technological applications, is equivalent in scope to that of molecular weight analysis in characterizing bulk polymers. Reviews of the various techniques and important areas of application of particle size analysis can be found in several references (e.g. (1,2)).

   In the past, analyses of submicron particles have been limited to time-consuming techniques, such as electron microscopy, or, to methods such as light scattering, which require a fairly narrow size distribution for accuracy. Recently, reports of a number of studies of a new method have been published in which modifications

0097–6156/84/0245–0003$06.50/0
© 1984 American Chemical Society

of the chromatographic techniques used in polymer molecular weight analysis have been employed to determine particle size and particle size distribution of suspensions (see reference (3) for a brief overview). These papers stem from experimental studies (4,5) which demonstrated that stabilized, dilute suspensions of latex particles fractionate by size when pumped through beds of porous or nonporous packing. There is now a clear indication that the development of such techniques for sizing submicron particles will have much the same impact on the science and technology of colloidal systems as liquid size exclusion chromatography analysis has had on the field of bulk polymers.

The purpose of this paper is to present a brief overview and description of a modelling approach we are taking which is aimed at developing a quantitative understanding of the mechanisms and separation capabilities of particle column chromatography. The main emphasis has been on the application of fundamental treatments of the convected motion and porous phase partitioning behavior of charged Brownian particles to the development of a mechanistic rate theory which can account for the unique size and electrochemical dependent separation behavior exhibited by such systems.

Background Description and Review of Separation Mechanisms

The experimental methods reported for particle chromatography have employed glass or stainless steel columns packed with nonporous copolymer or glass beads, porous gel matrices, or various GPC porous glass materials. Most studies have analyzed polymer latex solute particles suspended in stabilized aqueous media with the common mode of signal detection being light scattering. Small's work (4) with various nonporous packing systems demonstrated that for a range of eluant ionic strengths, larger latex particles elute from the column ahead of smaller ones and that the primary factors affecting the elution time were eluant ionic strength, packing diameter, and flow rate. The fractionation process occurs solely in the mobile phase and results from the fact that the latex particles are preferentially excluded from the slower moving solvent streamlines nearest the packing surfaces and thus obtain average velocities in excess of the solvent and these velocities increase with solute size. The name Hydrodynamic Chromatography or HDC has therefore been used to describe the process.

A number of publications (6-10) have demonstrated that the size separation mechanism in HDC can be described by the parallel capillary model for the bed interstices. The relevant expression for the separation factor, R_F, (ratio of eluant tracer to particle mean residence times) is given by,

$$R_F = \langle v_p \rangle / \langle v_m \rangle \qquad (1)$$

where the particle and marker average velocities through the bed, $\langle v_p \rangle$ and $\langle v_m \rangle$, are given by

$$\langle v_p \rangle = \frac{\displaystyle\int_0^{R_o - R_p} v_p(r) \exp\frac{-\phi(r)}{kT} \, r dr}{\displaystyle\int_0^{R_o - R_p} \exp\frac{-\phi(r)}{kT} \, r dr} \tag{2}$$

and (8)

$$\langle v_m \rangle = \frac{\displaystyle\int_0^{R_o} v_o\left(1 - \frac{r^2}{R_o^2}\right)\exp\left[\frac{-2e\psi_{01}}{kT}\exp(-\kappa a)\right] r dr}{\displaystyle\int_0^{R_o} \exp\left[\frac{-2e\psi_{01}}{kT}\exp(-\kappa a)\right] r dr} \tag{3}$$

In these equations, R_o represents the equivalent capillary radius (given by the bed hydraulic radius (7)), R_p is the particle radius, v_o is the eluant maximum velocity in the capillary tube, ψ_{01} is the packing surface potential, e is the protonic charge, κ is the inverse Debye double layer thickness, a is the distance of approach of the solute and wall, k is the Boltzmann constant, and T is the temperature. The expression for $v_p(r)$ in Equation 2 contains a correction for the hydrodynamic wall effect, and the total potential energy of interaction, ϕ, contains terms for the double layer and Born repulsion, and van der Waals attraction (8). The expression in Equation 3 is the limiting form appropriate for an ionic marker species (8).

Fits (in some cases zero free parameter (8,12)) of Equations 1 to 3 to experimental data have shown excellent argeement with the model (8-10,12) including an explanation of the ionic strength role of surfactants (10), universal calibration behavior (8,10), and the possibility of separating equi-sized particles of differing chemistry at high ionic strength conditions (3). The model therefore offers an excellent quantitative vehicle for describing the HDC mechanism. Of particular note is that the need for specification of the potential energy effects, ϕ, and hydrodynamic effects, $v_p(r)$, requires specification of a flow geometry. In this respect, modelling of particle chromatography is in some sense more restrictive in its assumptions than the psuedo-continuum rate theories which have been developed for macromolecular size exclusion chromatography (26-28).

The work of Krebs and Wunderlich (5) has been followed by a number of studies (3,13,15-20), demonstrating that particle size fractionation will also occur with a porous matrix. In this case,

in addition to the purely hydrodynamic effects, the possibility
exists for added fractionation due to steric exclusion from the
matrix pores, similar to the macromolecular size exclusion
mechanism (21). However, to our knowledge only one paper (13) has
given a quantitative model for a separation mechanism for porous
HDC. Calculations were based on a modified form of the
hydrodynamic model developed to describe the separation mechanism
of size exclusion chromatography (22-24). In this model the bed
is assumed to consist of a fraction, Φ_i, of capillaries of radius
R_o in a parallel array with a fraction, Φ_p, of flow-through
capillaries whose radius equals that of the packing pores. The
expression which results for the separation factor is (13)

$$\frac{1}{R_F} = \frac{\Phi_P}{R_{F,p}} + \frac{\Phi_i}{R_{F,i}} \qquad (4)$$

where $R_{F,p}$ and $R_{F,i}$ are respectively, the separation factors for
the porous matrix capillaries, and the interstitial capillaries,
given in each case by an expression in the form of Equations 1 to
3 (the upper limit radius R_o in this case refers to either the
porous matrix capillary radius or the interstitial capillary
radius).

Figure 1 shows the fit obtained using Equation 4 with the
appropriate expressions for the potential energy and wall effect
parameters and corrections for the micelle phase (13). The data
were obtained with a large pore diameter (2.5 μm) Fractosil
packing. One sees that the separation factor increases over that
for HDC despite the fact that larger packing size, in this case
90 μm, should lead to a reduction (4,8). This represents an
influence of the small pores in the separation behavior. Although
the model calculations can reasonably well describe the trend of
the data, the fit is not as convincing as the HDC model for
nonporous sytems (despite the fact that the interstitial capillary
radius and Hamaker constants have been slightly adjusted (13)).
On the other hand, model calculations varying parameters, show
clearly that the smaller diameter capillaries, representative of
the porous matrix, do play a controlling role in the separation
factor behavior.

The presence of the pores adds two parameters - the pore
volume fraction and the pore radius. The predicted R_F increases
as the pore radius decreases suggesting a preference for small
pore packings. However, for a small pore radius of 1.0 μm a
single value of the separation factor corresponds to two values of
the particle diameter (13). Such double-valued behavior is of
course undesirable in an analytic technique.

An obvious shortcoming of these calculations is that no
account is taken of the possibility of size exclusion phase
partitioning of the particle-pore system.

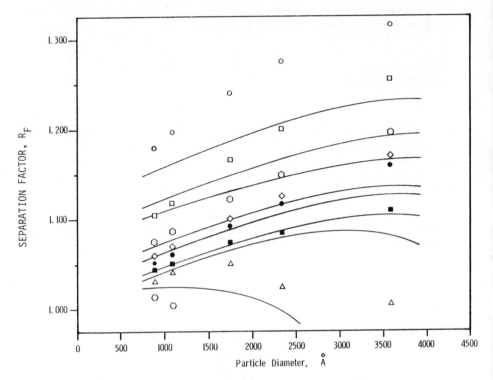

Figure 1.   Comparison between capillary HDC model calculations (-) and experimental data. (Reproduced with permission from Ref. 13.   Copyright 1980, Plenum Publishing Corporation.)   Total ionic strengths:

    ○ = 0.00022M   □ = 0.00055M   ◔ = 0.00129M
    ◇ = 0.00515M   ● = 0.0101M   ◼ = 0.0210M
    △ = 0.035M   ⬠ = 0.105M

Pore Partitioning HDC

The basis for Equations 2 and 4 has been a more or less ad hoc
comparison to the relations for the mean residence time rigorously
derived by Brenner and Gaydos (25) for particle transport through
small capillaries. In order to compare mechanistic models for
porous and nonporous chromatographic systems, a fundamental basis
for deriving separation factor expressions is needed since in
general both the HDC and pore-partitioning processes have to be
accounted for. A rigorous starting point for deriving a particle
chromatography rate theory would be the psuedo-continuum or volume
averaging process which has been used for classical chromatography
(see for example discussions in references (26 - 28)). In our
approach, we are taking the view that the details of the bed
geometry are needed in order to evaluate important hydrodynamic
wall and electrostatic interaction effects. However, in order to
arrive at a workable set of equations which can be tested against
experiment, a number of simplifications are necessary. For
example in HDC, use of the parallel capillary bed model in effect
means the volume averaging process is simplified to an area-
average of the transport equations across the bank of tubes
representing the bed. Employing a steady state assumption for the
radial concentration gradient leads directly to the applicable
expression (see reference (25) Equation 4.22 and arguments
preceeding) for the interstitial pore flow (29).

$$\frac{\partial C_m}{\partial t} + \langle v_p \rangle \frac{\partial C_m}{\partial Z} = \bar{D}^* \frac{\partial^2 C_m}{\partial Z^2} \tag{5}$$

In Equation 5, C_m represents the average or bulk concentration of
solute particles in the mobile phase, $\langle v_p \rangle$ is the average solute
particle velocity given in Equation 2 and $\bar{D}^*$ is the
phenomenological dispersion coefficient given in (25). Neglecting
the dispersion effect in Equation 5 leads, by means of the moment
analysis (25) directly to the HDC expression given in Equations 1
to 3.

In the case of porous HDC, as indicated, one needs to account
for both HDC, pore partitioning, and hindered diffusion
processes. A model should also have as asymptotes the mean
residence time behavior given by Equations 1 to 3 for a nonporous
system and Equation 4 for a purely flow-through porous system.
Rate equation analyses for classical size exclusion chromatography
have been based on treating the porous matrix as a homogeneous,
spherical medium within which radial diffusion of the
macromolecular solute takes place (e.g. (28,30,31)) or if mobile
phase lateral dispersion is considered important, a two
dimensional channel has been used as a model for the bed (32). In
either case, however, no treatment of the effects to be expected
with charged Brownian solute particles has been presented. As a

first approach to this problem we have carried out a simplified
analysis of the rate theory equation to be expected for a porous
system in which only partitioning occurs. The bed geometry is
assumed to consist of a series of parallel capillaries with
attached, cylindrical pores as shown in Figure 2.

The starting point is the convective-diffusion equation
suitably modified to account for wall effects and potential field
effects (25).

$$\frac{\partial P}{\partial t} + \underset{\sim}{\nabla} \cdot [P\underset{\sim}{v}_p - \underset{\sim}{D} \cdot \underset{\sim}{\nabla}P - P\underset{\sim}{M} \cdot \underset{\sim}{\nabla}\phi] = 0 \qquad (6)$$

In Equation 6, the diffusivity and mobility are second rank
tensors whose positional dependence is a consequence of the
hydrodynamic wall effect and P represents the probabililty that
the Brownian particle, initially at some fixed point, will be at
some position in space $\underset{\sim}{R}$ at a later time t. At low
concentrations, P is replaced by the number concentration, C
(25). Conceptually the approach followed is similar to that
developed by Brenner and Gaydos (25), however, one needs to
include an expression for the flux of particles at the wall due to
exchange with the pores. Upon averaging over the interstitial tube
cross section of Figure 2, one arrives at the following expression
(29) for the area averaged rate equation for the mobile phase
transport.

$$\frac{\partial C_m}{\partial t} + \langle v \rangle_p \frac{\partial C_m}{\partial Z} = \bar{D}_m \frac{\partial^2 C_m}{\partial Z^2} - \frac{(1-\Phi_i)\bar{D}_s}{\ell \Phi_i}\left(\frac{\partial C_s}{\partial x}\right)_{x=0} \qquad (7)$$

In Equation 7, solute concentrations are area averaged with the
subscripts referring to the mobile or stationary phase, ℓ is the
length of the dead end pores, and the diffusivities $\bar{D}_m$ and D_s
refer to the appropriately averaged values as defined in (25) to
account for hydrodynamic wall effects and potential energy
profiles between the packing and solute particles. Since in the
present analysis these will either be neglected or treated as
constants, rewriting their form in detail is not necessary.
Similarly for the stationary pore phase, one has for the
simplified one dimensional case,

$$\frac{\partial C_s}{\partial t} = \bar{D}_s \frac{\partial^2 C_s}{\partial x^2} \qquad (8)$$

Initial conditions and boundary conditions complete the model
description:

$$C_s = C_m = 0, \text{ for all z at t = 0} \qquad (9a)$$

$$C_m \text{ is bounded as } z \to \infty \qquad (9b)$$

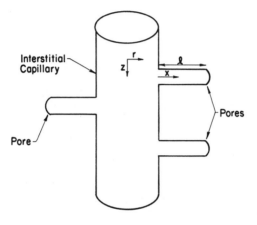

Figure 2. Schematic illustration of capillary-cylindrical
pore model for porous-partitioning HDC system.

$$\int_o^\infty [\Phi_i C_m + \Phi_p C_s] \, dz = M \qquad (9c)$$

$$\bar{D}_s \left(\frac{\partial C_s}{\partial x}\right)_{x = \ell} = 0 \qquad (9d)$$

$$C_s = KC_m \qquad (9e)$$

where M is the mass of solute particles injected per unit column area and K is the equilibrium partition coefficient (33,34).

Equation 9e expresses the assumption of local equilibrium of the partitioning process at the stationary phase – mobile phase interface.

Equating chemical potentials for the particle concentrations in the mobile and stationary phases leads directly to the expression for K in terms of the potential energy, ϕ, the particle radius, R_p, and the pore radius R (29,33-35).

$$K = \frac{\int_o^{R-R_p} \exp(-\phi/kT) \, r \, dr}{\int_o^R r \, dr} \qquad (10)$$

Evaluation of K is possible using the forms for sphere-plane interactions (11) (a simplification necessitated by the otherwise complicated forms needed to account for pore wall curvature (35)).

The solution of Equations 7 and 8 evaluated at the column exit yields the chromatogram. Since these equations cannot be solved analytically, statistical moments were obtained using the method of Laplace transforms (29).

The first moment is the mean retention time θ:

$$\theta = [1 + K\sigma] \, (L/\langle v_p \rangle) \qquad (11)$$

where L is the column length, and $\sigma = \dfrac{1-\Phi_i}{\Phi_i}$.

The second moment is the standard deviation, μ_2,

$$\mu_2 = 2/3 \, [K\sigma\ell^2/ \, \bar{D}_s](L/\langle v_p \rangle) \qquad (12)$$

In these expressions, it has been assumed that the mobile phase diffusivity will be negligible (29).

Equation 11 when manipulated according to the definitions of R_F yields an expression for the separation factor:

$$\frac{1}{R_F} = \frac{1}{R_{F,i}} \, [\frac{1 + K\sigma}{1 + K_m \sigma}] \qquad (13)$$

where K_m is the partition coefficient for the marker species, and $R_{F,i}$ is the separation factor in the interstitial capillaries as given by equations 1 to 3.

In the limit of total exclusion of both the marker and the particles from the pores ($K = K_m = 0$), the separation factor equals the separation factor in the interstitial capillaries. This is not true if the particles but not the marker ions are excluded ($K = 0$, $K_m \neq 0$). Also, in the limit of zero pore volume, the separation factor equals the separation factor in the interstices.

Interaction Energy Expressions. Previous papers (8,10,12,13) have used exact sphere-plane interaction energy expressions to approximate the sphere-cylinder interaction. In this work, these exact expressions were replaced with recently published approximate expressions. For the double layer repulsion, this avoided the inconvenience and inaccuracy of using tabular values (8) while for the van der Waals attraction, using the approximate solution simplified the programing task.

The previously mentioned expressions were originally derived by Bell et al. (36) to calculate the double layer repulsion. These expressions are valid for $\kappa R_p > 5$ where κ is the inverse Debeye length. For $\kappa R_p \leq 5$, tabular values (8) were used.

For our work, expressions of Ohshima et. al. (37) obtained from an approximate form of the Poisson-Boltzmann equation were used. These analytical expressions agree with the exact solution for $\kappa R_p \geq 1$. (All of our calculations meet this criterion.) The relation between the surface potential and the surface charge density is (37)

$$I = 2 \sinh (Ys/2) \left\{ 1 + \frac{2}{A \cosh^2 (Ys/4)} + \frac{8 \ln [\cosh (Ys/4)]}{A^2 \sinh^2 (Ys/2)} \right\}^{1/2} \quad (14)$$

where $A = \kappa R_p$, I is the dimensionless surface charge density, and Ys is the dimensionless surface potential defined in (37).

The double layer interaction energy is given in terms of the eluant dielectric constant ε by (37).

$$\phi_{DL} = \varepsilon (\frac{kT}{e})^2 R_p [4 \tanh (Y_{pk}/4)] Y_2 \exp (-\kappa a) \quad (15)$$

where $Y_2 = 8 \tanh(Ys/4) \left[\frac{1}{1 + \{1 - \frac{2A+1}{(A+1)^2} \tanh^2 (\frac{Ys}{4})\}^{1/2}} \right]$ (15b)

a is the distance of approach between the particle and wall surfaces, and Y_{pk} is the dimensionless packing potential.

Expressions used for the van der Waals energy were originally developed by Clayfield and Lumb (38) for the van der Waals attraction between a sphere and a flat plate. These complex expressions have a discontinuity in the first derivative at the transition between the region of small separations, for which the retardation effect is negligible, and the region of large separations for which retardation must be considered. In this work, an approximate expression developed by Gregory (39) was used:

$$\phi_{vw} = - \frac{A_H}{6a} \left(\frac{1}{1 + 14a/\lambda} \right) \tag{16}$$

where ϕ_{vw} is the interaction energy between a sphere and a flat plate, A_H is the Hamaker constant, and λ is the wavelength of the dispersion interaction, given in reference (8). This expression is valid for $\phi_{vw}/A_H < 0.1$. For values greater than 0.1, the van der Waals attraction is so small that any error will be insignificant.

The effects of experimental parameters on the predicted separation factor for the partition model are shown in Figures 3 to 6. As was seen with the parallel capillary model, the separation factor increases with particle diameter, increases with decreasing ionic strength, becomes more sensitive to the Hamaker constant as the ionic strength increases, increases with decreasing pore radius, and increases with decreasing packing size. In contrast to the parallel capillary model, the partition model predicts that the separation factor at low ionic strengths does not approach a constant value as the particle diameter increases.

Separation factors predicted by the partition model are compared with the experimental data from reference (13) in Figure 7. The partition model predicts the magnitude of the separation factor better than the parallel capillary model (see Figure 1), however the parallel capillary model predicts the shape of the curves better. This suggests that neither model alone is sufficient to account for the separation.

Combination of Models

The previous calculations indicate that both flow-through pore capillaries and partitioning may contribute to porous HDC separation. To investigate this possibility, the column can be modelled as banks containing large capillaries, small capillaries, and large capillaries with attached cylindrical pores as illustrated in Figure 8. In the model sketched, || refers to the portion of the bed cross-section which consists of a parallel array of large diameter flow-through interstitial capillaries, denoted ℓ, and small diameter flow through capillaries, s, which correspond to the portion of the porous phase which is

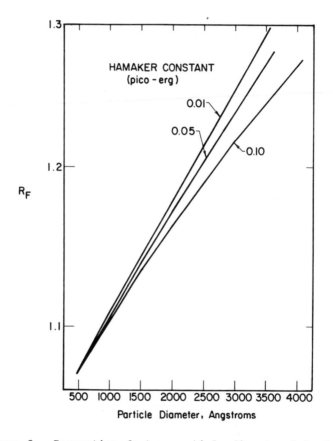

Figure 3. Separation factor-particle diameter behavior
computed from the pore-partitioning model showing the
effect of the Hamaker constant at a low eluant ionic
strength (0.001 M). Other parameters are Φ_i = 0.60, interstitial capillary radius = 16 μm, pore radius = 1.25 μm,
cylinder (packing) surface potential 30 mV, particle surface charge density = 1.5 x 10^4 stc/cm^2, ϵ = 74.3, and T
= 300 $^\circ$K.

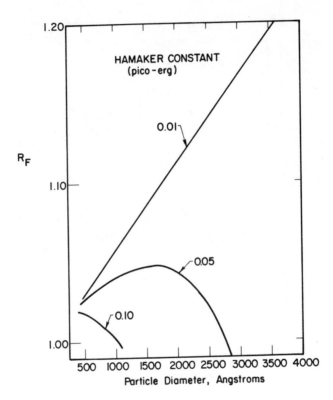

Figure 4. Separation factor-particle diameter behavior
computed from the pore partitioning model showing the
effect of the Hamaker constant at high ionic strength, 0.1
M. Other model parameters have same values as Figure 3.

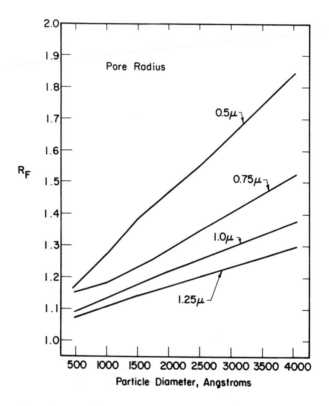

Figure 5.   Separation factor-particle diameter behavior as
a function of the pore radius for the pore-partioning model.
Hamaker constant = 0.05 pico-erg; all other parameters are
the same as in Figure 3.

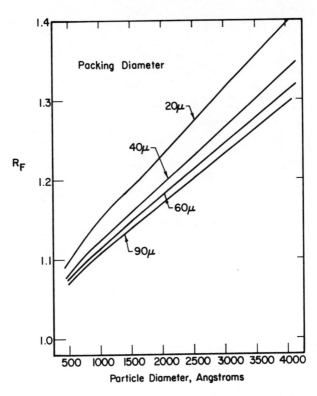

Figure 6. Separation factor-particle diameter behavior as a function of packing diameter for the pore-partitioning model. Parameters are the same as in Figure 3 with the exception of the interstitial capillary radius which was computed from the bed hydraulic radius (Equation 11 (7), with void fraction = 0.358).

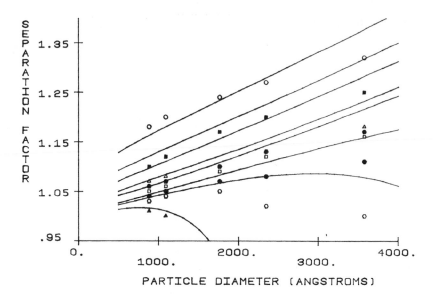

Figure 7. Comparison between R_F - particle size data of
Figure 1 and the pore partitioning model. Parameters for
model are the same as in Figure 3. Total ionic strengths:
    Total ionic strengths:
        ○ = 0.00022M   ▰ = 0.00055M   △ = 0.00129M
        ● = 0.00515M   ◻ = 0.0101M   ◖ = 0.0210M
        ◑ = 0.035M   ▲ = 0.105M

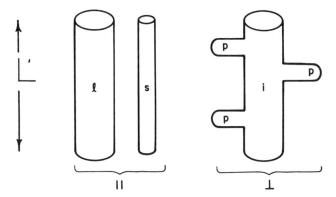

Figure 8. Schematic illustration of bed cross section for
the combination model. See text for explanation of nomenclature.

flow-through. The portion of the cross section associated with the partitioning process is denoted ⊥, and consists of large, flow-through interstitial capillaries, i, to which are attached stagnant pore volume cylinders, p, with which particles partition during passage through the i tubes. Derivation of the separation factor for this model follows the development given earlier for a purely flow-through system (13). The particle elution volume is given by

$$V_{R,P} = Q_F \langle t \rangle \tag{17}$$

where Q_F is the eluant flow rate, and $\langle t \rangle$ is the mean residence time. Since the average residence time is the sum of the times the particle spends in each capillary,

$$\langle t \rangle = n_\ell \langle t \rangle_\ell + n_s \langle t \rangle_s + n_\perp \langle t \rangle_\perp \tag{18}$$

where n_j is the total number of capillaries of type j the particle samples, and $\langle t \rangle_j$ is the average retention time in a capillary of type j.

Since the probability that a particle will sample a capillary of type j is given by

$$p_j = N_j q_j / Q_F \tag{19}$$

where N_j is the number of capillaries of type j in a bank, and q_j is the flow rate in a capillary of type j, then the total number of capillaries of type j a particle samples is

$$n_j = n p_j = n N_j q_j / Q_F \tag{20}$$

where n is the number of banks. The average marker velocity in a capillary of type j is given by

$$\langle v_m \rangle_j = q_j / a_j \tag{21}$$

where a_j is the cross-sectional area of the capillary and the average time for a particle in a given capillary is

$$\langle t \rangle_j = \frac{L'}{\langle v_p \rangle_j} \tag{22}$$

Combining Equations 17 to 22 yields

$$\frac{1}{R_F} = \frac{V_\ell}{V_m} \frac{1}{R_{F,\ell}} + \frac{V_s}{V_m} \frac{1}{R_{F,s}} + \frac{V_i}{V_m} \frac{1}{R_{F,\perp}} \frac{\langle v_m \rangle_i}{\langle v_m \rangle_\perp} \tag{23}$$

where V_j is the volume of all capillaries of type j, and $R_{F,j}$ is the separation factor in a capillary of type j.

The expression for the mean particle retention time (Equation 11) may be written for the marker and manipulated to yield

$$\frac{\langle v_m \rangle_\perp}{\langle v_m \rangle_i} = \frac{1}{1 + K_m \sigma} \tag{24}$$

Substituting Equations 13 and 24 into Equation 23 yields

$$\frac{1}{R_F} = \frac{V_\ell}{V_m} \frac{1}{R_{F,\ell}} + \frac{V_s}{V_m} \frac{1}{R_{F,s}} + \frac{V_i}{V_m} \frac{1}{R_{F,\ell}} [1 + K\sigma] \tag{25}$$

Since $R_{F,\ell}$ simply refers to the HDC separation factor associated with the interstitial capillaries (i and ℓ of Figure 9), to be consistent with the nomenclature of Equation 4 we shall now refer to it as $R_{F,i}$. Likewise $R_{F,s}$ refers to the hydrodynamic flow-through pores of Figure 8 and in the nomenclature of Equation 4 this becomes $R_{F,p}$. Making these changes in Equation 25 and rearranging gives the final result.

$$\frac{1}{R_F} = \frac{\Phi_i}{R_{F,i}} + \frac{(1 - \psi)(1 - \Phi_i)}{R_{F,p}} + \frac{\Psi(1 - \Phi_i)K}{R_{F,i}} \tag{26}$$

In Equation 26, the term Φ_i is as originally defined (i.e., volume fraction of the total void volume associated with the interstitial void space), ψ represents the fraction of the porous phase with which the particles interact by pure partitioning, and Ψ represents the fraction of the interstitial void volume associated with the partitioning process ($\perp$ phase of Figure 9). The definition of Φ_i is based on the total void volume associated with the marker species, which in general will not be the same as the true void volume due to electrostatic repulsion and partitioning of the marker. On the other hand, V_m is the experimentally accessible quantity and the problem formalism explicitly corrects for the latter effects through expressions of the type given by Equations 3 and 24.

It is useful to check the asymptotic behavior of this expression. For nonporous packing, $\Phi_i = 1$, and Equation 26 becomes

$$\frac{1}{R_F} = \frac{1}{R_{F,i}} \tag{27}$$

which is the expression for nonporous HDC.

For porous packing with all flow-through pores, $\psi = 0$, $\Psi = 0$, and Equation 26 becomes

$$\frac{1}{R_F} = \frac{\Phi_i}{R_{F,i}} + \frac{(1 - \Phi_i)}{R_{F,p}} \tag{28}$$

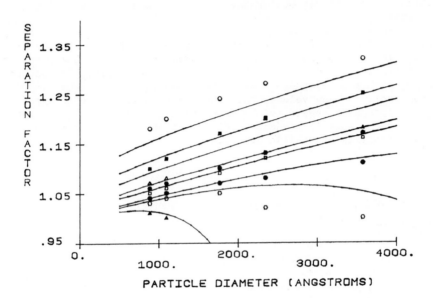

Figure 9.   Comparison between the data of Figure 1 and the combination model for $\psi = \Psi = 0.5$.   All other parameters are the same as in Figure 3.   Ionic strengths are the same as in Figure 8.

and since for this case, $\Phi_p = 1 - \Phi_i$ the above is the expression
for the parallel capillary model of porous HDC given by Equation 4.

For a system which consists of a porous matrix which is purely
partitioning, all of the interstitial capillaries will have
connected pores, $\Psi = 1$, and all of the pores will partition
particles, $\psi = 1$, thus Equation 26 becomes

$$\frac{1}{R_F} = \frac{\Phi_i}{R_{F,i}} (1 + K\sigma) \qquad (29)$$

which is the expression for the partition model with $K_m = 1$ given
in Equation 13.

Figure 9 shows the data fit obtained by use of Equation 26
with $\psi = \Psi = 0.5$. These results illustrate that by including all
three mechanisms (HDC in small and large capillaries, and particle
partitioning) an improved fit results. At this point, it must be
emphasized that although ψ and Ψ are computational parameters with
arbitrarily chosen values, they represent physically meaningful
quantities with regard to the separation process.

Further work is needed to relate all of the parameters to
experimentally accessible quantities.

Literature Cited

(1)   Collins, E. A.; Davidson, J. A.; Daniels, C. A. J. Paint
      Technol. 1975, 47, 604.
(2)   Kavanaugh, M.; Leckie, J. O., Eds.; "Particulates in
      Water: Characterization, Fate, Effects, and Removal";
      ADVANCES IN CHEMISTRY SERIES, American Chemical Society:
      Washington, D. C., 1980.
(3)   McHugh, A. J.; Nagy, D. J.; Silebi, C. A. in "Size Exclusion
      Chromatography (GPC)"; Provder, T., Ed.; ACS SYMPOSIUM
      SERIES No. 138, American Chemical Society: Washington,
      D.C., 1980; pp. 1-25.
(4)   Small, H. J. Colloid Interface Sci., 1974, 48, 147.
(5)   Krebs, K. F.; Wunderlich, W. Angew. Makromol. Chem., 1971,
      20, 203.
(6)   Stoisits, R. F.; Poehlein, G. W.; Vanderhoff, J. W. J.
      Colloid Interface Sci. 1976, 6, 237.
(7)   McHugh, A. J.; Silebi, C. A.; Poehlein, G. W.; Vanderhoff,
      J. W. "Colloid Interface Science"; Vol. IV, Academic: New
      York, 1976, p. 549.
(8)   Silebi, C. A.; McHugh, A. J. AIChE J. 1978, 24, 204.
(9)   Prieve, D. C.; Hoysan, P. M. J. Colloid Interface Sci.
      1978, 64, 201.
(10) Nagy, D. J.; Silebi, C. A.; McHugh, A. J. J. Colloid
      Interface Sci. 1981, 79, 264.
(11) Silebi, C. A.; McHugh, A. J. J. Appl. Polym. Sci. 1979, 23,
      1699.

(12) Silebi, C. A.; McHugh, A. J. in "Emulsions, Latices, and Dispersions"; Becher, P.; Yudenfreund, M. N., Eds.; Marcel Dekker: New York, 1978; pp. 155-174.

(13) Nagy, D. J.; Silebi, C. A.; McHugh, A. J. in "Polymer Colloids II"; Fitch, R. M., Ed.; Plenum: New York, 1980; pp. 121-137.

(14) Nagy, D. J.; Silebi, C. A.; McHugh, A. J. J. Appl. Polym. 1981, 26, 1555.

(15) Gaylor, V. F.; James, H. L. Preprints, Pittsburgh Conference on Analytical Chemistry, Cleveland, March 1975.

(16) Coll, H.; Fague, G. R. J. Colloid Interface Sci. 1980, 76, 116.

(17) Singh, S.; Hamielic, A. E. J. Appl. Polym. Sci. 1978, 22, 577.

(18) Nagy, D. J.; Silebi, C. A.; McHugh, A. J. J. Appl. Polym. Sci. 1981, 26, 1567.

(19) Johnston, J. E.; Cowherd, C. L.; MacRury, T. B. in Size Exlusion Chromatography (GPC)", T. Provder, Ed., ACS SYMPOSIUM SERIES, No. 183, American Chemical Society; Washington, D.C., 1980, pp. 27-45.

(20) Husain, A.; Hamielec, A. E.; Vlachopoulous, J. ibid, pp. 47-75.

(21) Yau, W. W.; Kirkland, J. J.; Bly, D. D. "Modern Size-Exclusion Liquid Chromatography"; Wiley: New York; 1979.

(22) DiMarzio, E. A.; Guttman, C. M.; Macromolecules, 1970, 3, 131.

(23) Guttman, C. M.; DiMarzio, E. A.; Macromolecules, 1970, 3, 681.

(24) Verhoff, H. F.; Sylvester, N. D.; Macromol. Sci-Chem., 1970, A4, 979.

(25) Brenner, H.; Gaydos, L. J.; J. Colloid Interface Sci., 1977 58, 312.

(26) Lee, H. L.; Lightfoot, E. N.; Reis, J. F. G.; Waissbluth, M. D. in "Recent Developments in Separation Science",; Li, N. N.; Ed. CRC Press: Cleveland; 1977, pp. 1-69.

(27) Novak, J.; Janak, J.; Wicar, S.; in Jour. of Chromatogr., 3, "Liquid Column Chromatography", Deyl, Z.; Macek, K.; Janak, J.; Eds. Elsevier: New York; 1975; pp. 25-43.

(28) Horn, F. J. M. AIChE J., 1971, 17, 613.

(29) Francis, D. C., M. S. Thesis, University of Illinois, 1983.

(30) Hermans, J. J., J. Polym. Sci., 1968, A-2, 6, 1217.

(31) Reis, J.F.G.; Lightfoot, E. N.; Noble, P. T.; Chang, A. S. Sep. Sci. Techn., 1979, 14, 367.

(32) Ouano, A. C.; Barker, J. A. Sep. Sci., 1973, 8, 673.

(33) Giddings, J. C.; Kucera, E.; Russell, C. P.; Myers, M. N. J. Phys. Chem., 1968, 72, 4397.

(34) Giddings, J. C. in "Treatise Anal. Chem.", 2nd ed., P. Elving, Ed., 1982, 1, (5).

(35) Smith, F. G.; Deen, W. M. J. Colloid Interface Sci., 1980, 78, 444.

(36) Bell, G. M.; Levine, S.; McCartney, L. N. J. Colloid
      Interface Sci., 1970, 33, 335.
(37) Ohshima, H.; Healy, T. W.; White, L. R. J. Colloid Interface
      Sci., 1982, 90, (1), 17.
(38) Clayfield, E. J.; Lumb, E. C. Disc. Faraday Soc., 1966, 42,
      285.
(39) Gregory, J. J. Colloid Interface Sci., 1981, 83, (1), 138.

RECEIVED September 12, 1983

Computer Model for Gel Permeation Chromatography of Polymers

DONG HYUN KIM[1] and A. F. JOHNSON[2]

Department of Chemical Engineering, University of Waterloo, Waterloo, Ontario,
Canada, N2L 3G1

A novel dynamic mass-balance model has been developed
to describe the fractionation of polymers by gel permeation chromatography. The model embodies several
dimensionless parameter groups which are particularly
convenient to use in order to predict the performance
of the chromatograph under a wide variety of conditions. It is shown that the molecular separation
processes are readily explained in terms of the
accessible void volume fraction in the gel column
packing material and the broadening effect by a
dimensionless parameter (α) which is a function of
column length, particle radius of the column packing,
eluant flow rate and the effective diffusivity of
the polymer molecules in the gel. Good agreement
has been observed between the model predictions and
experimental results. The model predictions are
compared with other published data.

Since the technique was introduced ($\underline{1}$) in 1964, gel permeation
chromatography (GPC), or size exclusion chromatography (SEC), has
played an increasingly important role for the characterisation of
polymers. The theory and practice of this chromatographic method
have been extensively reported and a comprehensive text has recently been published on modern size exclusion chromatography ($\underline{2}$).
   One of the least well understood aspects of the whole field
is the precise physical nature of the process whereby polymer
chains of a different size **are** separated by passage through a gel
column. On a qualitative level adequate explanations of the phenomenon exist but it has proved to be a more difficult task to
formulate and solve anything other than the simplest of mathematical models of the chromatographic process.
   Broadly, there are two classes of model which have found
application ($\underline{2}$): The plate theory is based on an oversimplified

[1]Current address: Korea Advanced Institute of Science and Technology, Seoul, Korea
[2]Current address: School of Polymer Science, University of Bradford, Bradford, BD7 1DP,
England

0097-6156/84/0245-0025$06.25/0

view of the chromatographic process and has been widely described
and used especially as an aid to the interpretation of gas chromatographic phenomena. The appeal of the plate theory is its simplicity but, as will be seen later, in many aspects it is far from
adequate when used to describe some experimentally observed phenomena in gel permeation chromatography. The alternative modelling
approach, the rate theory, appears to have grown out of the van
Deemter equation (3) and takes into account axial dispersion phenomena and mass transfer between the bulk flow and the column
packing material. A limitation of the van Deemter equation is that
it does not take into account intra-particulate diffusion.
However, an extension of this approach through the use of Fick's
Law, has made it possible to formulate the differential equations
which describe solute mass balance in a very small column section.

There have been relatively few applications of the rate theory
to GPC, presumably because of the apparent complexity of this
approach. One of the most widely quoted interpretations of the
rate theory to GPC is that of Ouano and Baker (4). These authors
have attempted to take advantage of the undoubted potential of the
rate theory approach in constructing a model. They identified the
key parameters in their model as the flow rate of the eluant, gel
particle size, diffusion coefficient in the stationary and mobile
phases and the partition coefficient for solute between phases.
Although there is little doubt that the important parameters have
been correctly identified, it is not immediately apparent how they
are inter-related and hence how their coupled effect can be interpreted. A critical account of the various attempts which have been
made to model the GPC process will be given elsewhere.

In the model described in this work every effort has been made
to ensure that it embodies physically meaningful parameters. It is
inevitable, however, that some simplistic idealizations of the
physical processes involved in GPC must be made in order to arrive
at a system of equations which lends itself to mathematical solution. The parameters considered are, the axial dispersion, interstitial volume fraction, flow rate, gel particle size, column
length, intra-particle diffusivity, accessible pore volume fraction
and mass transfer between the bulk solution and the gel particles.
A coherent inter-relationship has been established between each of
these parameters through a few, readily handled, dimensionless
parameters.

Amongst the assumptions we have made in developing the model
are the following: that Fick's law is applicable to the diffusion
processes, the gel particles are isotropic and behave as hard
spheres, the flow rate is uniform throughout the bed, the dispersion in the column can be approximated by the use of an axial dispersion coefficient and that polymer molecules have an independent
existence (i.e. very dilute solution conditions exist within the
column). Our approach borrows extensively many of the concepts
which have been developed to interpret the behaviour of packed bed
tubular reactors (5).

Model Development

The proposed model can be developed by consideration of three important steps in the chromatographic process:

(i)   Dispersion and/or backmixing.
(ii) Mass transfer between the gel and the mobile phase.
(iii) Diffusion of the solute within the gel structure.

These are illustrated in Figure 1. The importance of these phenomena can be best shown by a general description of the gel permeation separation process.

The introduction of the polymer sample solution to the chromatographic column can be regarded as a sharp concentration pulse and is usually commonly represented mathematically as a Dirac delta function. Although this is an adequate description of the concentration pulse, it does not adequately represent the polydispersity which might exist in the polymer. In all modelling studies the polymer sample (e.g. monodisperse polystyrene standards), have been considered to be truly monodisperse although it is known that they do have a Poisson distribution of molecular sizes (6). As the sample is eluted through the packed column it is fractionated according to molecular size by the difference in the accessible pore volume of the gel. The chromatogram will be 'broadened' by a combination of factors such as diffusion within the gel particles, dispersion in the mobile phase and solute transfer between the gel and the eluant. The observed chromatogram is, in effect, the sum of many overlapping peaks.

The gel particles making up the column packing are commonly spherical in shape and have diameters in the range of $5-50 \times 10^{-6}$ m. In order to facilitate modelling it is assumed that in any given column the particles are all of equal size and that each particle has an equal pore-size distribution. It is generally accepted that, when column packing particles have a diameter which is a factor of at least 20 less than that of the column, plug flow with some superimposed dispersion can be assumed in the column.

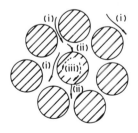

Figure 1. Three different mechanisms for fractionation:
(i) Dispersion; (ii) Mass transfer between gel and mobile phase; (iii) Diffusion within the gel structure.

On the above basis the mobile phase material balance equation can be written as:

$$D_{ax} \frac{\partial^2 c^*}{\partial X^2} - u \frac{\partial c^*}{\partial X} - \frac{3 D_e}{R_0} \left(\frac{1 - \varepsilon}{\varepsilon}\right) \left(\frac{\partial c_s^*}{\partial r^*}\right)_{R_0} = \frac{\partial c^*}{\partial t} \qquad (1)$$

where D_{ax} is the axial diffusion coefficient, D_e the effective diffusivity coefficient for the solute in the gel, c^* is the solute concentration in the mobile phase, c_s^* solute concentration within the gel, R_0 the radius of the gel particle, X and r^* represent distance variables along the column and within the particle respectively and ε is the interstitial volume fraction.

The stationary phase material balance equation is:

$$\frac{D_e}{\varepsilon_p} \left(\frac{\partial^2 c_s^*}{\partial r^{*2}} + \frac{2}{r^2} \frac{\partial c_s^*}{\partial r^*}\right) = \frac{\partial c_s^*}{\partial t} \qquad (2)$$

Here ε_p is the accessible pore volume fraction of the gel which is a function of the pore size distribution as well as the size of a polymer molecule. The value of ε_p is given by Equation 3:

$$\varepsilon_p = \int_{\bar{r}}^{\infty} K_{SEC} (\bar{r}, \eta) \, \Phi (\eta) \, d\eta \, / \int_{\bar{r}}^{\infty} \Phi (\eta) \, d\eta \qquad (3)$$

where K_{SEC} is the solute distribution coefficient, $\bar{r}$ the equivalent hard-sphere radius, η the pore radius and Φ is the pore size distribution.

The initial and boundary conditions are:

$$c^* (x, o) = c_s^* (r^*, 0) = 0 \qquad (4)$$

$$K_f (c^* - c_s^*) = \frac{D_e}{\varepsilon_p} \left(\frac{\partial c_s^*}{\partial r^*}\right) \quad \text{(for all X, } r^* \leq R_0) \qquad (5)$$

$$c_{in}^* (t) = c^* (o^+, t) - \frac{D_{ax}}{u} \frac{\partial c^*}{\partial X} \Big|_{X = o^+} \qquad (6)$$

$$\frac{\partial c^*}{\partial X} \Big|_{X = L^-} = 0 \qquad (7)$$

Equations 6 and 7 are the well known Dankwerts boundary conditions (7). In Equation 5, K_f is the mass transfer coefficient around the gel, and c_{in}^* (t) the inlet solute concentration which is a function of analysis time.

The above equations can be most conveniently handled in their dimensionless forms.

$$\frac{1}{Pe} \frac{\partial^2 c}{\partial z^2} - \frac{\partial c}{\partial z} - 3\alpha \left(\frac{1-\varepsilon}{\varepsilon}\right) \left(\frac{\partial c_s}{\partial r}\right)_{r=1} = \frac{\partial c}{\partial \theta} \tag{1a}$$

$$\frac{\alpha}{\varepsilon_p} \left[\frac{\partial^2 c_s}{\partial r^2} + \frac{2}{r} \frac{\partial c_s}{\partial r}\right] = \frac{\partial c_s}{\partial \theta} \tag{2a}$$

where $\quad \alpha = \dfrac{L\, D_e}{u\, R_o^2}$, $\quad z = \dfrac{X}{L}$, $\quad \theta = \dfrac{ut}{L}$, $\quad Pe = \dfrac{Lu}{D_{ax}}$, $\quad r = \dfrac{r^*}{R_o}$ $\qquad$ (3a)

$$c = \frac{C^*}{C_o} , \quad c_s = \frac{C_s^*}{C_o}$$

The equivalent dimensionless initial and boundary conditions are:

$$c\,(z,\,o) = c_s\,(r,\,o) = o \tag{4a}$$

$$\gamma\,(c - c_s) = \left(\frac{\partial c_s}{\partial r}\right) \quad \text{at } r = 1,\ \text{all } z \tag{5a}$$

$$c_{in}\,(\tau) = c\,(o^+,\,\tau) - \frac{1}{Pe} \frac{\partial c}{\partial z} \bigg|_{z\,=\,o^+} \tag{6a}$$

$$\frac{1}{Pe} \cdot \frac{\partial c}{\partial z} \bigg|_{z=1^-} = o \tag{7a}$$

where $\quad c_{in} = \dfrac{C_{in}^*}{C_o}$, $\quad \gamma = K_f \dfrac{R_o\, \varepsilon_p}{D_e}$

Equations 1a - 7a do not have an analytical solution in a closed form in the time domain and of necessity have to be handled in the Laplace domain.

The Laplace domain solution is:

$$\frac{\overline{c}\,(z,\,s)}{\overline{c}_{in}\,(s)} = 2\,\exp\left(\frac{Pe \cdot z}{2}\right) \frac{\sinh\left[\frac{Pe}{2}\,\beta\,(1-z)\right] + \beta\,\cosh\left[\frac{Pe}{2}\,\beta\,(1-z)\right]}{(1+\beta^2)\,\sinh\left(\frac{Pe}{2}\cdot\beta\right) + \cosh\left(\frac{Pe}{2}\cdot\beta\right)} \tag{8}$$

where $\quad \beta = \sqrt{1 + \dfrac{4}{Pe}\,h\,(s)} \qquad$ (9)

and $\quad h(s) = 3\alpha\left(\dfrac{1-\varepsilon}{\varepsilon}\right) \cdot \gamma \cdot \dfrac{\left[\sqrt{\dfrac{s\cdot\varepsilon_p}{\alpha}} \coth\sqrt{\dfrac{s\cdot\varepsilon_p}{\alpha}} - 1\right]}{\left[\sqrt{\dfrac{s\cdot\varepsilon_p}{\alpha}} \coth\sqrt{\dfrac{s\cdot\varepsilon_p}{\alpha}} - 1 + \gamma\right]} \qquad$ (10)

Here s is the Laplace variable and $\overline{c}_{in}$ (s) the Laplace transform of the input. When $\overline{c}_{in}$ (t) can be approximated by a Dirac delta function, $\overline{c}_{in}$ (s) = 1 and the right hand side of Equation 8 is the Laplace transform of the solute concentration at any z.

The solutions of Equations 8-10 were obtained with an IBM 370 computer using an improved version of the Filon method (8).

The leading moments at the exit (z-1) were used to obtain the mean, variance and skewness of the peaks and these can be calculated from the relation:

$$M_n = \int_0^\infty \theta^n \, c(1, \theta) \, d\theta = \lim_{s\to 0} (-1)^n \frac{d^n}{ds^n} \overline{c} (1, s) \tag{11}$$

The first moment M_1 provides the mean dimensionless elution time:

$$M_1 = \delta_1 - 1 + \frac{1 - \epsilon}{\epsilon} \cdot \epsilon_p \tag{12}$$

The second moment is:

$$M_2 = \left(\frac{2}{Pe}\right)^2 \delta_1^2 \left(\frac{Pe^2}{4} + \frac{Pe}{2} - \frac{1}{2} + \frac{1}{2} \exp(-Pe)\right) + \delta_2 \tag{13}$$

$$\text{where} \quad \delta_2 = \frac{2}{3} \alpha \left[\frac{1 - \epsilon}{\epsilon}\right] \left(\frac{\epsilon_p}{\alpha}\right)^2 \left[\frac{1}{5} + \frac{1}{\gamma}\right] \tag{14}$$

and the third moment is:

$$M_3 = \left(\frac{2}{Pe}\right)^3 \delta_1^3 \left(-3 + \frac{3}{4} Pe + \frac{3}{4} Pe^2 + 3 \exp(-Pe) + \frac{Pe^3}{8} + \frac{9}{4} Pe \exp\right.$$
$$(-Pe)) + 3\delta_1\delta_2 \left(\frac{2}{Pe}\right)^2 \left(\frac{Pe^2}{4} + \frac{Pe}{2} - \frac{1}{2} + \frac{1}{2} \exp(-Pe)\right) + \delta_3 \tag{15}$$

$$\text{where} \quad \delta_3 = \frac{2}{3} \alpha \left(\frac{1 - \epsilon}{\epsilon}\right) \left(\frac{\epsilon_p}{\alpha}\right)^3 \left(\frac{6}{105} + \frac{2}{5\gamma} + \frac{1}{\gamma^2}\right) \tag{16}$$

The central moments may be obtained from Equation 17:

$$\mu_n = \int_0^\infty (\theta - M_1)^n \, c(1, \theta) \, d\theta \tag{17}$$

The second central moment which provides the variance of the distribution has the form:

$$\mu_2 = \left(\frac{2}{Pe}\right)^2 \delta_1^2 \left(\frac{Pe}{2} - \frac{1}{2} + \frac{1}{2} \exp(-Pe)\right) + \delta_2 \tag{18}$$

The third central moment is:

$$\mu_3 = 3 \; (\frac{2}{Pe})^3 \; \delta_1^3 \; (\frac{Pe}{2} - 2 + (\frac{Pe}{2} + 1) \; \exp \; (-Pe)) + 3\delta_1\delta_2$$

$$(\frac{2}{Pe})^2 \; (\frac{Pe}{2} - \frac{1}{2} + \frac{1}{2} \exp \; (-Pe)) + \delta_3 \tag{19}$$

The peak skew is given by:

$$\mu_s \;\; = \;\; \frac{\mu_3}{\mu_2^{3/2}} \tag{20}$$

From the general plate theory (2) it is known that:

$$L \; . \; \frac{\sigma^2}{V_R^2} \; = \; H \tag{21}$$

where H is the individual plate height, σ^2 the variance and V_R the mean retention volume. Therefore it follows that:

$$H \; = \; L \; (\frac{\mu_2}{M_1^2}) \tag{22}$$

$$= \; L \; (\frac{2}{Pe})^2 \; (\frac{Pe}{2} - \frac{1}{2} + \frac{1}{2} \exp \; (-Pe)) + \frac{\delta_2}{\delta_1^2} \tag{23}$$

Column efficiency is most simply related to L/H.

Model Input Parameters
_____

In solving the model the quality of the result depends greatly on the accurate estimation of many parameters. It is not always easy to estimate the parameter required by the proposed model, hence some attention will be given to the methods we have adopted in obtaining them.

(i) Accessible Pore Volume Fraction, ε_p. One of the primary factors in the effective separation of polymer molecules according to size is the accessible volume of the gel pores which is a function of the solute size as can be clearly seen from Equation 3.

Both K_{SEC} and the pore size distribution have been measured experimentally for hard-sphere column packing materials (9), but for soft gel packing materials there does not seem to be any reliable information presumably because the accepted method of pore structure characterisation in porous materials, mercury porosimetry, cannot be used. However, ε_p can be measured for gels without great difficulty from the column calibration curve (as is manifest from Equation 12) provided the calibration is made on the basis of the peak mean position, i.e. the first moment of the peak

rather than the usual peak maximum. Obviously for perfectly symmetrical peaks the mean and the maximum coincide. In this work we
have adopted the latter route to ϵ_p.

(ii) Effective Diffusivity, D_e. Effective diffusivity in porous
structures is relatively well understood for gases but much less
well understood for liquids and is virtually unknown for polymer
solute molecules (10). In attempting to arrive at a meaningful
estimate for D_e we have adopted the following simple model (11)
(Equation 24) which has frequently been used for gas phase effective diffusivity.

$$D_e = \frac{D_m \, \epsilon_p \, K_r}{\tau} \tag{24}$$

In this equation D_m is the diffusivity of the polymer solute in
the bulk solution which may be estimated (12) from Equation 25:

$$D_m = \frac{RT}{6\pi\mu_o N_o} \left[\frac{10\pi \, N_o}{3K}\right]^{1/3} \left(\overline{M}_v\right)^{-(1+a)/3} \tag{25}$$

where R is the gas constant, T the absolute temperature, μ_o the
solvent viscosity K and a are the constants of the Mark-Houwink-
Sakura equation and $\overline{M}_v$ the viscosity average molecular weight.
In Equation 24, K_r is the fractional reduction in the diffusivity
within the gel pores which may be attributed to the friction effects of the solute with the 'walls' of the pores and it, in turn,
may be obtained (11) through the use of Equation 26.

$$K_r = (1 - 2.104 \, \lambda + 2.09 \, \lambda^3 - 0.95 \, \lambda^5) \tag{26}$$

where $\lambda = \overline{r}/\overline{\eta}$ (27)

The value of $\overline{\eta}$ was estimated using the simple approximation:

$$\overline{\eta} = (\overline{r} + \eta_{max})/2 \tag{28}$$

The η_{max} can be obtained from the GPC calibration curve by estimating the maximum molecular size of the solute which can penetrate
the pores of the gel.
      To be more rigorous K_r should be obtained from Equation 29:

$$K_r = \int_r^\infty (1 - 2.104\lambda + 2.09\lambda^3 - 0.95\lambda^5) \, \Phi \, (\eta).d\eta \Big/ \int_\gamma^\infty \Phi \, (\eta) \, d\eta \tag{29}$$

but since the use of this equation requires a value for the pore
size distribution in the gel it cannot be readily used. As far as
we are aware there are no experimental data available for pore
size distributions in soft gels.

In Equation 24, τ is the tortuosity, a term well established for gas diffusion into porous materials (10). It is unfortunate, but necessary to introduce τ into our model. The value of τ cannot be obtained a-priori and must be obtained experimentally since it is an almost impossible task to describe the complicated pore geometry in a gel. Given idealised perfect pore geometry it has been possible to estimate τ for gas diffusion processes. In our work, of necessity τ becomes an adjustable parameter to help achieve better agreement between the model predictions and experimental results. Since τ is unmeasurable our only concern has been to use reasonable values in our simulations.

It will be seen later that it is very important to use a 'good' value for D_e in order to obtain agreement between the model predictions and experimental chromatograms. The parameter D_e is not only responsible for the fractionation of the polymer but also in determining the extent of broadening.

(iii) Mass Transfer Coefficient, K_f. Under the normal operating conditions for a chromatograph, particularly for the more recent high performance instruments, the Reynolds number, (Re_p) is very low ($\sim 0.001 - 0.5$). Numerous correlations have been proposed for K_f for situations where Re_p is relatively large (13), but for GPC only one appears to be suitable (14) for the estimation of K_f (Equation 30).

$$\epsilon j_D = 1.09 \ Re_p^{-2/3} \qquad 0.0016 < Re_p < 55 \qquad (30)$$

where j_D is the mass transfer Colburn j factor ($\frac{K_f}{G} \rho Sc^{2/3}$) G being the mass velocity of fluid, ρ is the fluid density and Sc the Schmidt number.

The value of K_f estimated in this way is far greater than the magnitude of D_e and for this reason plays a negligible role in the overall broadening effect as will be seen later. As already mentioned, the van Deemter model is also inadequate under most operating conditions of chromatographs as it only takes into account external (the gel particles) mass transfer and neglects the internal diffusion of the solute through the gel structure.

(iv) Axial Dispersion, D_{ax}. There are ample descriptions of the axial dispersion phenomenon in the field of packed bed reactors (15). The axial dispersion coefficient embodies all the factors which contribute to broadening from inter-particulate movement of solution. In this work D_{ax} can be related to terms of interest through Equation 31 since

$$\epsilon Pe_p = 0.011 \ Re_p^{0.48} + 0.20 \qquad (31)$$

$$Pe_p = \frac{2\bar{u}R_o}{D_{ax}} \qquad (32)$$

It is apparent from Equation 31 that when Re_p is very low, then Pe_p has an essentially constant value.

Experimental

A Waters Associates GPC (Model ALC/GPC 301) was used for experimental measurements. The instrument was fitted with a 100 μl injection loop, a UV detector and a single 10^3 A Styragel column. The column specifications are:

Catalogue No. 26913 (Waters Associates)
Particle size 37-75 μ
Plate count   350
Column ID   7.8 mm
Column length 61 cm

The following polystyrene standards (Pressure Chemical Co.) were used:

Sample	$\overline{Mn}$	$\overline{M}_w/\overline{M}_n$
A	1,800,000	1.28
B	390,000	1.09
C	51,000	1.05
D	37,500	1.05
E	17,500	1.05
F	9,000	1.05
G	4,000	1.10
H	2,100	1.15
Toluene	92	1.0

All measurements were carried out in THF at ambient temperature with sample concentrations of 0.1 wt. %. Normally a flow rate of 1.25 ml min^{-1} was used but with sample F further measurements were made at a flow rate of 2.21 ml min^{-1}.

Results and Discussions

The quality of the mathematical model can only be judged by its ability to predict the likely experimental results over a wide range of conditions. The goodness of the agreement between the predictions of our proposed model and experimental observations is very much dependent on the key parameters in the model being clearly defined and well understood as mentioned previously. One of the difficulties we have encountered in attempting to compare the model behaviour and experimental GPC traces has been in obtaining reliable estimates of Pe and D_e.

Calibration Curve. Figure 2 shows the calibration curve obtained for the single column. It is similar in character to others which have been reported.

This curve was used to calculate ε_p from Equation 12 (which in turn required the calculations of ε which will be described in the next section). It is important to note that the accessible pore volume varies for different solutes. If K_{SEC} and Φ (M) were explicitly available then Equation 3 could be used to obtain ε_p.

The Case of Total Exclusion. One would expect for very fine particles that in the case of polymer samples which are totally excluded from the particle pores that the observed chromatogram would be symmetrical as the Pe should be very high ($\sim$ 10,000). Figure 3 shows the experimentally observed result for a polymer sample of molecular weight $\sim$ 2,000,000. An identical result was obtained with a sample of molecular weight $\sim$ 390,000. It can be seen that the curve is much as expected showing only slight skewing. Since the two samples gave indistinguishable results we conclude that the very slight skewing which can be detected stems from the fact that the column did not behave precisely according to the assumptions we have made in this work where perfect symmetry would be expected. Obviously this kind of minor discrepancy cannot be readily quantified as so many factors could be used by way of explanation for the effect and these might well have to be changed from column to column.

Figure 3 also illustrates that good agreement between the model prediction and experimental observation was obtained when Pe = 1130. The value for Pe was low on the basis of the correlation equation, Equation 31, which predicts a value of $\sim$ 5000. Perhaps the discrepancy in Pe values is not too surprising in view of the large scatter of the data from which Equation 31 was obtained.

The total exclusion chromatogram provides the means to obtain the ε values and this was found to be 0.423. It is interesting to compare this value with that reported (4) for the interstitial volume of randomly packed rigid spheres which is 0.364. We assume that our value deviates from the hard sphere value primarily because of the inefficient packing of particles in the case of the column used in this work varied substantially in size (35 - 75 μ).

Chromatographic Curves.

(i) Symmetrical peaks - symmetrical chromatograms (within experimental error) were obtained with polymer samples of low molecular weight (< 9000). Our model predicts this symmetry as will become apparent when the importance of the dimensionless group α in the model is discussed below. It will be seen that as α increases then the resulting peak should approach perfect symmetry. For all low molecular weight samples, it is evident from Equation 25 that D_m must be relatively large and in consequence D_e must also be

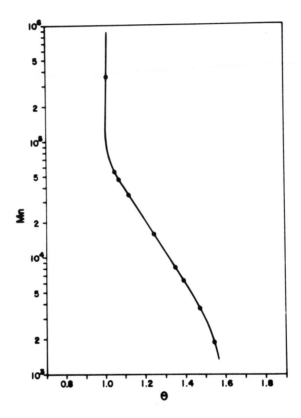

Figure 2. Calibration curve: this curve was obtained by using the mean of the chromatographic peak of each poly-styrene standard.

large (Equation 24) which increases α. The model readily fits
any symmetrical chromatogram as can be seen in the typical example shown in Figure 4. In fitting the symmetrical chromatograms
the calculated Pe (1130) was used throughout but some changes were
made to τ in order to achieve better fits. The τ values are normally in the range 2-10 for hard spheres with gas phase systems
(10). In this work values in the range 8-12 were adopted which
we feel are reasonable for this convenient 'general purpose' adjustable parameter.

The computer model is capable of dealing with any eluant
flow rate but it has only been possible to test the capacity of
the model over a very small range of flow rates because of the
danger of damaging a column when conducting experiments at high
flow rates. Figure 4 also shows the good agreement which was
obtained between experiment and calculation when the flow rate is
approximately doubled. Increasing the flow rate broadens the
peak and shifts the peak maximum in the manner anticipated (16).

(ii) Skewed peaks - for polymer of molecular weight 37,000 the
chromatographic peak was distinctly skewed. The model proposed
was also capable of fitting such curves as can be seen from Figure 5. To fit such curves the same Pe was used but the τ was
adjusted as before to achieve good fit between the model predictions and the experimental value.

On the basis of these preliminary experimental results we
are confident the proposed model is capable of explaining many
of the experimentally observed features of size exclusion chromatograms. It is perhaps appropriate to comment further on the
physical importance of the major parameters of interest in the
model.

The value of $K_f R_o \epsilon_p/D_e$ (or γ) will, in our experience be
large and on the order of several hundred. When Equations 13 and
15 are considered in the light of this information they can be
simplified and γ plays no part in the overall broadening effect.
This was readily apparent from the computer simulations.

The influence of Pe on the computed chromatogram can be seen
in Figure 6. When the Pe > 10,000 its influence on peak broadening is relatively insignificant. However, at low values it
plays a significant role in determining peak shapes. In the experimental work reported here the column used was very short and
hence Pe was also low and therefore influenced the peak shapes in
the simulations. When Re_p is small then Pe_p becomes constant
(see Equation 31) and hence Pe is proportional to column length.
If longer column lengths are adopted than those used in this work
(as will normally be the case) then Pe rapidly approaches a value
where its effect on broadening becomes negligible.

Following the above rationalisation it becomes apparent that
α is the only significant dimensionless group in our model. Its
influence is vividly demonstrated in Figure 7 which shows that by
altering α alone it is possible to cover all known peak shapes.

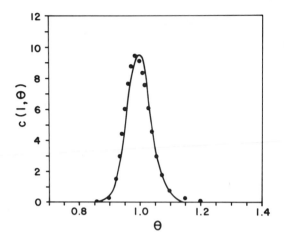

Figure 3. Model fitting of chromatogram of totally excluded standard polymer; $\overline{M}_n$ = 1,800000, u = 0.103 cm sec^{-1},
Pe = 1130, ε = 0.423, O = experimental data, ——— computed
curve.

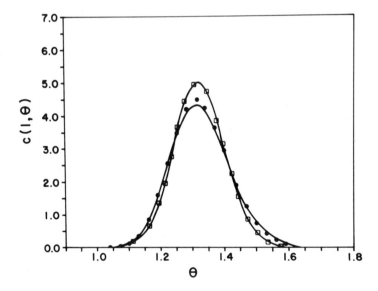

Figure 4. Model fitting of chromatograms which are approximately symmetrical; $\overline{M}_n$ = 9,000, Pe = 1130, ε = 0.423,
ε_p = 0.224, D_e = 3.79 x 10^{-8} cm^2 sec^{-1}. Experimental data:
u = 0.103 cm sec^{-1} ( ), u = 0.182 cm sec^{-1} ( ), ——— com-
puted curves with τ = 12.

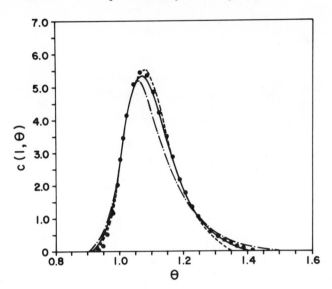

Figure 5. Model fittings of skewed chromatograms; $\overline{M}_n$ = 37,000, Pe = 1130, ε = 0.423, ε_p = 0.077, D_e = 2.55 x 10^{-9} cm^2 sec^{-1} (τ = 8), u = 0.103 cm sec^{-1}. Experimental data: ( ), ——— computed curves. The computed curves with τ = 6 (----) and τ = 12 (- . - . -).

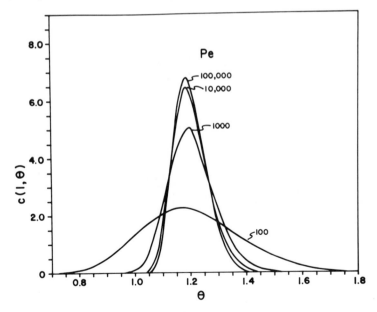

Figure 6. Effect of Pe values on model predictions. In each case α = 1.0, ε_p = 0.15, γ = 400, ε = 0.423.

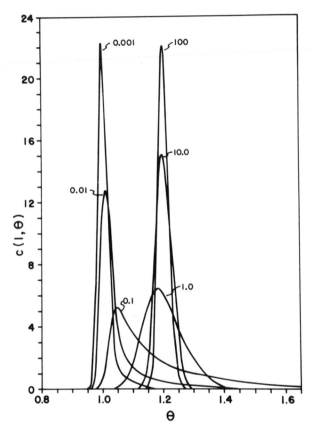

Figure 7. Effect of α values on model predictions. In each case Pe = 10,000, ε_p = 0.15, γ = 400, ε = 0.423.

When α is large (> 100) then the peak is symmetrical with a mean value of 1.20. In the extreme as $\alpha \to \infty$ then the simulated peak approaches the input function which in our model simulation is a Dirac delta function at $\theta = 1.2$. In the real situation it should approach the true MWD of the polymer being analysed. As α decreases the simulated peak is first broadened and then skewed. It is apparent that the peak maximum shifts in a manner expected. As $\alpha \to 0$ the peak becomes similar to that expected for total exclusion. There is an apparent anomaly evident from Figure 7. The simulated peaks appear to have very different means, particularly where α is small, yet one would expect from Equation 12 for the mean to be invariate with α. In fact, at low α values there are always extremely long tails at towards high θ which cannot be adequately shown pictorially. The mean does, in fact, remain constant and it is only when $\alpha = 0.0$ that the peak mean abruptly shifts to 1.0, i.e. there is a discontinuity of the mean as a function of α at zero. This can be verified by the manipulation of Equations 8-11.

It is evident from Figure 7 that as α increases the resolution as well as the symmetry is enhanced which explains the trend with modern columns towards fine gel particles. The parameter α is such that it leads to the conclusion that when achieving the same resolving power it is possible to decrease the total analysis time by reducing the particle size, the time of analysis being proportional to $1/R_0^2$. This prediction is substantiated by published work (see Figure 5) (17). In going from particle sizes of 120 µm to 6 µm it has been shown that the total analysis time is reduced from 250 min to 37.5 sec which conforms exactly to the proportionality factor suggested (17). By increasing u one would expect poorer chromatograph performance which can be seen from Figure 5. The performance of the GPC should also increase in direct proportion to column length. The resolving power should increase with increasing D_e (i.e. decreasing molecular weight). Each of the other predictions conform to well established observations.

The proposed model can be readily related to the plate theory. The number of theoretical plates can be deduced from the moment expressions and when Pe and the K_f are large then Equation 33 follows from Equation 23.

$$\frac{1}{N} = \frac{2}{Pe} + \frac{2}{15} \frac{(\frac{1-\epsilon}{\epsilon}) \frac{\epsilon_p^2}{\alpha}}{(1 + \frac{1-\epsilon}{\epsilon} \epsilon_p)^2} \tag{34}$$

It is evident that α also plays an important role in this equation. For a given ϵ_p and Pe as α increases then N also increases. This can be best demonstrated by varying the flow rate. Several experimental studies have been reported (18-19) with flow rates being changed over the range $\sim$ 0-8 ml/min. There is an interesting contradiction in these reports on the way in

which N varies at low flow rates. In one case (18) when using
gel particles of 5 μm and 10 μm a maximum is observed in N at
∿ 1-3 ml/min which suggests that there is an optimum operating
flow rate for GPC with these packing materials. In the same publication when 20 μm particles were used N increased continuously
down to very low flow rates, behaviour similar to that observed
by other workers (19) with 10 μm particles. The van Deemter
model was invoked to explain the observed maximum in N. However,
this model only takes into account the external mass transfer
effect and does not consider D_e. Our model suggests that the correct relationship between N and flow rate does not contain a maximum (see Figure 4). At most one might expect, at extremely high
α values a plateau at very low flow rates (0.1 ml min^{-1}) and
therefore the previously inexplicable experimental observations
are the correct ones according to our model. The experimentally
observed maximum may result from agglomeration effects with particles less than 20 μm in size or other experimental anomalies.
Clearly, additional experimental studies are desirable in order
to clarify the behaviour of N with flow rate.

A further interesting feature of plate count predictions
from the model is that N values can be compared with those quoted
by column manufacturers for their products. In this work the
quoted N value was 350 and that calculated was 280 at 1.25 ml/min
when using polystyrene standard $\overline{M}_n$ = 9000, i.e. one which gives a
symmetrical curve. In making such comparisons one must bear in
mind that the plate theory assumes that N is independent of polymer molecular weight which is not the case with our model. When
a standard of $\overline{M}_n$ = 37,000 was used the calculated N was 160.

It has been suggested (20) that N varies with $1/R_O$. Clearly,
this does not conform to our model prediction of N being proportional to $1/R_O^2$. We feel that this discrepancy stems entirely
from the limited range of R_O values which were used in the experimental work. Had a wide range been used the parabolic nature of
the relationship would probably have been seen.

On the basis of the above discussion we feel that our simple
dimensionless parameter appraoch has a generality which has advantages over previously described models (2).

There are many facets of this study which we feel merit further investigation. In particular it is necessary to consider an
extension of the proposed model, which in its present form is con-
fined to the performance of a simple column, to cover the behaviour of any set of columns since it is column sets which are
normally used. In addition, it is important to consider the input to the model which should be truly representative of polymers
with a molecular weight distribution and not merely a concentration pulse of perfectly monodisperse polymer. In relation to
this latter suggestion it would be significant if it were possible to link this model to the very real problem of deconvolution,
i.e. the removal of instrumental and column broadening from the
observed chromatogram to produce the true molecular weight distri-

bution of the input sample. Although we have given some preliminary consideration to this point, it is not immediately apparent
as to how the problem can be resolved mathematically.

Summary

Despite the assumptions and simplifications we have made in arri-
ving at a model we feel that the physical basis we have adopted
is sufficiently realistic to give good predictions, certainly as
far as our present experimental results enable us to make tests.
The numerical solution of the model equations we have used presented no difficulties using a fast computer ($\sim$ 5 secs per solution).
    We feel that one of the attractions of the approach we have
adopted is that very few parameters are required to control the
model. We have proposed that the use of a particularly important
dimensionless parameter α which plays a significant role not only
in predicting our own experimental observations but also in explaining the results of others. Indeed, several anomalies in the
literature have come to light as a result of comparing our predictions with published information.
    It is apparent that the plate theory cannot possibly explain
skewed chromatograms. We suggest that the major reason for this
limitation is that it is not correct to assume that an equilibrium is achieved between the solute in the mobile and stationary
phase during analysis.

Acknowledgments

The authors wish to express their gratitude to their respective
Institutions for leave of absence and to the Department of Chemical Engineering, University of Waterloo for its generosity in
providing the support and facilities which has made this work possible. We also appreciate the assistance given by Mr. A. Burcyzk
with GPC measurements and Mr. S. Lee for the drawing of the figures.

Glossary

α	$L.D_e/u.R_o^2$ dimensionless number
a	Mark-Houwink-Sakurada constant
D_{ax}	Axial diffusion coefficient
D_e	Effective diffusivity coefficient of the solute in the gel
D_m	Diffusivity of the solute in solution
$c = c^*/C_o$	Dimensionless concentration
c_o	Initial concentration of solute in eluant
c^*	Solute concentration in the mobile phase
$c_s = c_s^*/c_o$	Dimensionless concentration within gel
c_s^*	Solute concentration within the gel
$c_{in}^*(t)$	Inlet solute concentration which is a function of time

ε	Interstitial volume fraction
ε_p	Accessible pore volume fraction of gel
G	Mass velocity of fluid
H	Plate height
$j_D = K_f \cdot \rho \cdot Sc^{2/3}/G$	Colburn mass transfer j factor
K	Mark-Houwink-Sakurada constant
K_f	Mass transfer coefficient around gel
K_r	Fractional reduction in diffusivity within gel pores resulting from frictional effects
K_{SEC}	Solute distribution coefficient
μ_o	Solvent viscosity
μ_n	nth central moment
μ_s	Peak skewness
M_n	nth leading moment
$\bar{M}_v$	Viscosity average molecular weight
N	Number of theoretical plates
$Pe = L \cdot u/D_{ax}$	Dimensionless number
$Pe_p = 2\bar{u}R_o/D_{ax}$	Dimensionless number
R	Gas constant
R_o	Radius of gel particle
Re_p	Reynolds Number
$r = r^*/R$	Dimensionless radius
$\bar{r}$	Equivalent hard sphere radius
r^*	Distance within gel particle
s	Laplace variable
Sc	Schmidt Number
σ^2	Variance
T	Absolute temperature
τ	Tortuosity
η	Pore radius
$\bar{\eta}$	Mean pore radius
Φ	Pore size distribution
ρ	Density
$\theta = u \cdot t/L$	Dimensionless number
V_R	Mean retention volume
X	Distance along column
$z = X/L$	Dimensionless distance

Literature Cited

1. Moore, J. C.; J. Polym. Sci., PtA, 1964, 2, 835.
2. Yau, W. W.; Kirkland, J. J. and Bly, D. D.; 'Modern Size Exclusion Chromatography, Wiley-Interscience, N.Y., 1979.
3. van Deemter, J. J.; Zuiderweg, F. J.; Klinkenberg, A. Chem. Eng. Sci., 1956, 5, 271.
4. Ouano, A. C.; Barker, J. A. Sep. Sci., 1973, 8, 673.
   Ouano, A. C. Adv. Chrom., 1977, 15, 233.
5. Wen, C. Y.; Fan, L. T.; 'Models for Flow Systems and Chemical Reactors', Chemical Processing and Engineering, Vol. 3, L. F. Albright; R. N. Maddox; J. J. McKetta; Dekker Inc: NY, 1975.

6. Szwarc, M.; 'Carbanions, Living Polymers and Electron Transfer Processes', Interscience, 1968.
7. Danckwerts, P. V.; Chem. Eng. Sci , 1953, 2, 1.
8. Kim, D.H. and Chang, K. S. (to be published).
9. de Vries, A. J.; LePage, M.; Beau, R.; Guillemin, C. L., Anal. Chem., 1967, 39, 935.
10. Satterfield, C. N.; Sherwood, T. K.; 'The Role of Diffusion in Catalysis', Addison-Wesley, London, 1963.
11. Satterfield, C. N.; Colton, C. K.; Pitcher, W. H.(Jr.); A.I.Ch.E. Journal, 1973, 19(3), 628.
12. Rudin, A.; Johnston, H. K.; J. Polym. Sci., PtB, 1971, 9, 55.
13. Wakao, N.; Funazukuri, T.; Chem. Eng. Sci., 1978, 33, 1375.
14. Wilson, E. J.; Geankopolis, J.; Ind. Eng. Chem. Fund., 1966, 5, 9.
15. Chung, S. F.; Wen, C. Y.; A.I.Ch.E. Journal, 1968, 14, 857.
16. van Kreveld, M. E.; van den Hoed, N.; J. Chromatogr. 1978, 149, 71.
17. Otacka, E. P.; Acc. Chem. Res., 1973, 6, 348.
18. Vivilecchia, R. V.; Lightbody, B. G.; Thimot, N. Z.; Quinn, H. M.; J. Chromatogr. Sci., 1977, 15, 424.
19. Mori, S.; J. Appl. Polym. Sci., 1977, 21, 1921.
20. Dawkins, J. V.; Stone, T.; Yeadon, G.; Polymer, 1977, 18, 1179.

RECEIVED September 1, 1983

Pressure-Programmed Controlled-Flow Supercritical Fluid Chromatograph

E. W. ALBAUGH and D. BORST

Gulf Research & Development Company, Pittsburgh, PA 15230

P. TALARICO

Waters Associates, Milford, MA 01757

Supercritical fluid chromatography is a form of chromatography in which the system is held near the critical temperature of the mobile phase and pressure utilized to effect solvency and hence migration. The advantages of this technique have been shown to be increased mass transfer and the migration of high molecular compounds. Most of the instruments designed for this technique have not attempted to control the flow as pressure is programmed. In this paper, an instrument is described in which the inlet liquid flow is held constant and the pressure regulated by a pneumatically activated flow control valve at the exit of the column. This approach permits the use of a wide pressure program with a controlled flow and the use of several conventional liquid chromatographic detectors. Separations of model systems including normal aliphatic hydrocarbons, polynuclear aromatics and polymers with molecular weights ranging up to one million are reported.

Supercritical fluid chromatography is a form of chromatography in which the temperature is held near the critical temperature of the mobile phase and pressure utilized to effect solvency and hence migration. The advantages of this technique have been shown to be increased mass transfer and the migration of high molecular weight compounds.[1,2,3,4] Since this work was reported, high performance liquid chromatography has made rapid advancement and overshadowed much of the early appeal of supercritical fluid chromatography.[5] However, in the area of wide molecular weight-range samples, supercritical fluid chromatography with pressure programming appears to have advantages.[4] Jentoft has demonstrated the potential of this technique and described the design of a pressure programmed instrument.[6] In this instrument the system pressure

0097-6156/84/0245-0047$06.00/0

was controlled by programming the inlet pressure, but the flow
was not controlled. Bartman, in an instrument designed for use
with carbon dioxide, has used a flow meter and a motor driven expansion valve at the column exit to regulate pressure and gas
flow. (7) (8) The current state of the field has been reviewed by (10)
Randal, (8) Gere, (9) and Peaden.

In this paper, an instrument is described in which the inlet
liquid flow rate is held constant and the pressure regulated by a
pneumatically actuated flow control valve at the exit of the
column. This approach permits the use of a wide-range pressure
program with a controlled flow. Also, by selecting mobile phases
that are liquids at ambient laboratory conditions, several types
of conventional liquid chromatographic detectors may be utilized.

EXPERIMENTAL

A schematic diagram of the instrument is shown in Figure 1. The
liquid mobile phase flows from the reservoir, through a heated
chamber for degassing, and a 10 μ filter to a syringe pump (Ruska,
Cat. No. 1441 with a Boston Ratiotrol variable speed motor control). A safety relief line leads from the pump through a 207
bar rupture disc in the reservoir. A line from the pump also runs
to a pneumatic pressure transmitter (Moore Model No. 1735) which
provides the process (pressure) signal for a controller (pressure)
(Moore, Nullmatic Controller, Model 50). From the pump the solvent flows to shut-off valves A and B (High-Pressure Equipment Co.
Model 15-12 AF1-316). With B closed and A open, the solvent flows
through the sample valve (Chromatronix Model HPSV, with 25 μl
loop) and into the column oven. The sample valve is enclosed in
an oven with a maximum temperature of 200PC which is maintained
by a temperature controller (West Guardsman, Jr.). When B is open
and A is closed, the solvent flows through the preheater and into
the oven. The preheater consists of 2 ft of 1/8 in. stainless
steel tubing wound around a Chromalox heater and coated with 2 in.
of insulation. The temperature is maintained 10°C above the
column oven temperature.

Inside the column oven, the solvent flows through 0.75 m of
0.009 in. I.D. conditioning coil, through a low dead-volume tee
containing a thermocouple to monitor solvent temperature, and
then to the column. The column oven, with a 425°C maximum temperature, is heated by two 2-kilowatt wire wound heaters which are
controlled with a Gulton Model 2GB Controller which provides
either isothermal or programmed temperature control.

After leaving the oven, the mobile phase flows through 1 m
of 0.009 in. capillary tubing which is immersed in a heat exchanger to return the solvent to ambient temperature. A control
valve, (Research Control Valve, Precision Products, Tulsa, Okla-
homa, Type 78S with a P-9 trim) with a low dead-volume head as
shown in Figure 2, is placed after the heat exchanger. This valve
is positioned by the controller (pressure). The set point of the

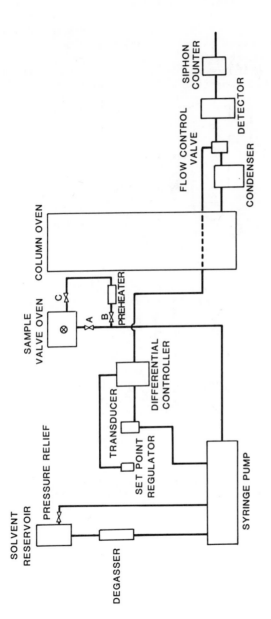

Figure 1. Supercritical fluid chromatograph.

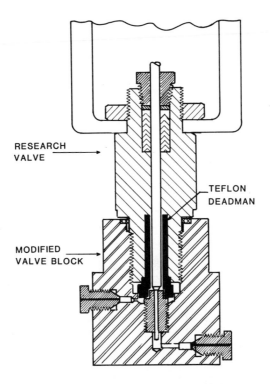

Figure 2. Modified research valve.

controller (pressure) is positioned by a variable speed motor
which is regulated by a GKHT-2 motor controller (G. K. Keller
Corp.). For isobaric operation, the set point is driven to the
desired position and the drive deactivated. The controller then
maintains the selected pressure. For pressure programming, the
motor speed is selected to drive the set point at the desired
rate. In operation, the flow rate is first set by the pump control and then the pressure adjusted. A siphon counter (Waters
Associates, Model C908) is placed at the end of the system to
measure volumetric flow.

The solvent system used here consisted of cyclohexane and 5%
ethanol. This mobile phase will dissolve many petroleum fractions, produce a stable base line, and is compatible under ambient
laboratory conditions with many common liquid chromatographic detectors. The system was operated at 10°C above the critical temperature of cyclohexane (280°C).

The stationary phase utilized was 75-100 mesh Porosil C
packed into four, 4-ft lengths of 1/4 in. O.D. stainless steel
tubing fitted with 10 μ snubber, swagelock 1/4 in. to 1/16 in.
unions.

The minimum pressure in this system at 1 ml/min. flow rate
and 280°C is 20.6 bar. The maximum operating pressure of the
instrument is 206.8 bar. As the pressure is increased, the flow
rate is constant from 20.6 to 48.2 bar and the base line is excellent. In the region from approximately 48.2-55.2 bar the flow
rate slows slightly and the base line rises with an ultraviolet
detector. This change is reproducible and believed due to a
phase change in the solvent system. From approximately 55.2 to
206.8 bar the flow rate is again constant and the base line excellent. Depending upon the pressure program rate, a short period of compression is initially required, and then the flow sta-
bilizes.

RESULTS AND DISCUSSION

The separation of a mixture of aromatic compounds (benzene, naphthalene, anthracene, chrysenes, and benz(a)pyrene) at 31 bar is
shown in Figure 3. This chromatogram was obtained with a Perkin
Elmer Model 250 ultraviolet detector with the high-pressure cell
placed after the cooling heat exchanger and before the flow control valve. A similar chromatogram is obtained with an Isco
Model UA4 with a 10 mm micro cell placed after the flow control
valve.

The effect of pressure (measured at the pump) on this separation can be seen in Figure 3 and Table I. As the pressure is increased, the retention volume of benzene and naphthalene continually increase, while the retention volume of anthracene, chrysene, and benz(a)pyrene first increase, go through a maximum, and
then decrease. The maximum separation occurs at 38.5 bar while
the minimum satisfactory separation volume (shortest analysis

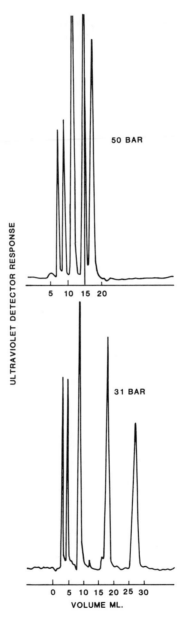

Figure 3.   Separation of benzene, naphthalene, anthracene, chrysene, and benz(α)pyrene at 50 and 31 bar.

time) occurs at 50 bar. At 51.7 bar, the higher molecular weight materials are not resolved.

Table I.   EFFECT OF PRESSURE ON SEPARATION OF
POLYNUCLEAR AROMATIC COMPOUNDS

			Elution Volume, ml			
Compound Pressure, bar	31	40	44.1	46.5	50	51.7
Benzene	2.5	5	6.0	6.5	8.0	8.0
Naphthalene	4.0	8	8.5	9.0	10.0	10.0
Anthracene	8.5	13.5	15.0	13.0	13.0	10-13
Chrysene	18.0	26.0	24.5	16.5	17	10-13
Benz(a)Pyrene	27.5	35.5	32.0	18.0	18.5	10-13

The separation of a somewhat higher molecular weight material is shown in Figure 4. Here, a polystyrene with an average molecular weight of 600 has been separated into eleven components at 50 bar. The effect of pressure programming is also shown. The program was started at 46 bar at a rate of 0.34 bar per min. As the pressure increases, the elution volumes of the various compounds decrease and the peak widths become more narrow in a manner similar to that for temperature programming in gas chromatography and solvent programming in liquid chromatography. At 65.5 bar the sample elutes essentially as one peak.

To demonstrate the behavior of high molecular weight compounds in this system, a series of polystyrene standards were analyzed. The highest molecular weight material (average molecular weight of 1,800,000) is shown in Figure 5. A pressure program rate of 0.69 bar per minute was used. A small amount of material elutes at approximately 75.8 bar, but the major portion of the sample elutes between 89.6 bar and 134.4 bar. A sample was taken from the 117.2 bar region and analyzed by conventional exclusion chromatography. It was found to have a molecular weight in the range of 1,000,000. Thus, these high molecular compounds survive the column and are resolved at pressures below 2000 psi. The other lower molecular weight polystyrene standards eluted at correspondingly lower pressures.

A series of normal hydrocarbons were analyzed using the same chromatographic system and the Pye LCM II flame ionization detector. In Figure 6 is shown the separation of a mixture of C_{16}, C_{32}, C_{40} and C_{44} normal hydrocarbons. With the high molecular weight capability shown for the polystyrenes, this system should also handle the higher molecular weight saturated hydrocarbons that are beyond the range of gas chromatography.

Several polynuclear compounds containing both aromatic and alkyl functions were chromatographed. The higher the molecular weight of the compound, the greater was the elution volume, indicating that separation was not occurring according to the number of aromatic rings.

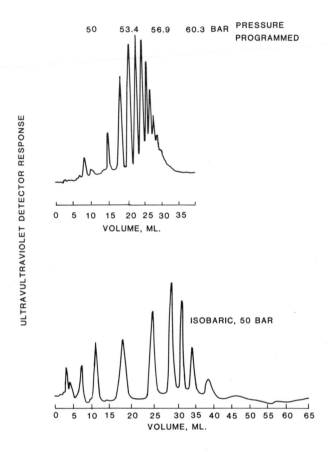

Figure 4. Separation of low molecular polystyrene by pressure programming.

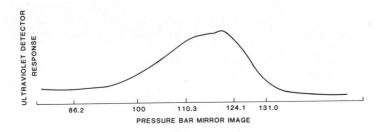

Figure 5. Pressure separation of 1,800,000 molecular weight polystyrene by pressure programming.

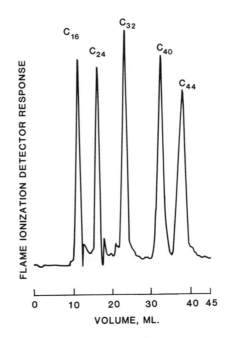

Figure 6. Separation of C_{16}, C_{24}, C_{32}, C_{40}, and C_{44} normal hydrocarbons at 48.3 bar mirror.

In the area of polar compounds, phenol, resorcinol, and benzoic acid were chromatographed. Phenol gives a nearly symmetrical peak with essentially no tailing. Resorcinol is separated from phenol but shows some tailing. Benzoic acid falls in the same elution region as resorcinol, but tails to a much greater extent.

In one application, styrene still bottoms were chromatographed. Here styrene and the individual lower molecular weight oligomers were separated, and as the pressure was increased, the higher molecular weight polystyrene eluted.

The instrument described here has been found to be essentially trouble-free. Pressure settings and control are reproducible, requiring only the positioning of a switch. One of the attractive attributes of supercritical fluid chromatography is the short time required for the column to reach equilibrium when conditions are changed. After operating at 172.4 bar, for example, the instrument can be rapidly depressurized to 20.7 bar and the system, including the columns, equilibriated within a few minutes. The column conditioning problems often found in high-pressure liquid chromatography-solvent programming were not experienced here. The reproducibility of elution volumes is comparable to isocratic high pressure liquid chromatography.

LITERATURE CITED

1. Giddings, J. C., Myers, M. N., and King, J. W., J. Chromatogr. Sci., 7, 276(1969).
2. Giddings, J. C., Science, 162, 7(1968).
3. Su, S. T., Rynders, G. W. A., Anal. Chem. Act., 38 31(1967).
4. Gouw, T. H., Jentoft, R. E., J. Chromatogr., 68, 303(1972).
5. Doran, T., Soc. Analyt. Chem., 117, May 1974.
6. Jentoft, R. E., Gouw, T. H., J. Chromatogr. Sci., 8, 138(1970).
7. Bartman, D., Berichte der Bunsen--Gesellschaft Bd. 76, NY 3/4, 336(1972).
8. Randal, L. G., Separation Science and Technology, 17(1) 1(1982).
9. Gere, D. R., Board, R., McManigill, Anal. Chem., 54, 736(1982).
10. Peaden, P. A., Lee, M. L., J. Liq. Chromatogr. 5(2), 179(1982).

RECEIVED October 13, 1983

Automated Data Analysis System for a Gel Permeation Chromatograph with Multiple Detectors

M. E. KOEHLER, A. F. KAH, T. F. NIEMANN, C. KUO, and T. PROVDER

Glidden Coatings and Resins, Division of SCM Corporation, Strongsville, OH 44136

A Waters Model 150C ALC/GPC was interfaced to a
minicomputer system by means of a microcomputer for
automated data collection and analysis. Programs
were developed for conventional molecular weight
distribution analysis of the data and for liquid
chromatographic quantitative composition analysis
of oligomeric materials. Capability has been
provided to utilize non-standard detectors such as
a continuous viscometer detector and spectroscopic
detectors for compositional analysis. The
automation of the instrument has resulted in
greater manpower efficiency and improved record
keeping.

Efficient use of a modern high performance gel permeation
chromatography (HPGPC) instrument requires computer aided
analysis in order to take full advantage of both the quality and
the quantity of information the instrument is capable of
providing. Commercial data analysis packages for this purpose
are, for the most part, simplistic and inflexible. This is
particularly true when multiple or non-standard detectors are
required. This work describes an automated data analysis system
used in conjunction with a Waters Associates Model 150C ALC/GPC
to read operational parameters from the instrument, to collect
data from multiple detectors, and transmit the data to a
minicomputer system for storage, analysis, reporting and
plotting.

Data Acquisition System

Automated data analysis for the chromatograph is achieved by
interfacing the instrument and detectors to a microcomputer for
data acquisition. The microcomputer is connected to the Intelink

0097-6156/84/0245-0057$06.00/0

interface of the instrument so that the operational parameters
for each sample analysis can be transfered to the minicomputer.
The microcomputer is responsible for all real-time activities
involved in data collection. At the completion of the
experiment, data are transferred via a serial line to the
minicomputer for storage and analysis. Report generation and
plotting may be done at any time after the completion of the
experiment. The minicomputer system uses a Digital PDP 11/44
processor running the RSX 11-M operating system. Programming for
communications and data analysis on the minicomputer is done in
FORTRAN-77. The microcomputer uses an 8080A processor and is
composed primarily of standard Pro-Log circuit cards.
Programming for the microcomputer is done in assembly language
and cross assembled on the minicomputer. Details of the
mini-microcomputer system and its organization have been reported
elsewhere (1,2,3).

Automated Instrument Analysis Process

There are four stages in an automated instrument analysis. In
the first stage, the instrument operator initiates the experiment
by means of dialog programs on the minicomputer. Examples of the
dialogs for the HPGPC operation are shown in Figures 1-4.
        Dialog 15, shown in Figure 1, is used for sample definition.
This includes identification of the location of the sample in the
automatic injector, the column set in use, the data collection
rate, the detectors to be used, the operators initials and the
sample identification. This definition file may be modified and
displayed on the terminal, or printed. The file is updated
during operation to show the current status of the samples.
Before initiating an analysis, the instrument must be programmed
for automatic operation and the samples placed in the appropriate
positions of the injector. Dialog 16, shown in Figure 2,
starts operation of the microcomputer. Intelink communication
with the instrument is established and the parameters for the
first sample are taken from the sample definition file on the
minicomputer and are transmitted to the microcomputer. The
microcomputer turns on a ready status light at the instrument
to signal to the operator to begin automatic operation of the
instrument.
        The second stage is data acquisition. This stage is entered
when the operator starts the instrument. The instrument makes
the first injection and signals the microcomputer via Intelink.
After a delay proportional to the void volume of the column set,
data are collected on a time basis (constant flow rate assumed)
at the predetermined rate from each of the detectors selected, up
to a maximum of three simultaneous detectors. When the sample
run is complete, the instrument again signals the microcomputer
which places the instrument in a hold state while it reads the
operational parameters from the instrument for that sample and

```
DIA 15

Instrument No. 34 - HPGPC Sample Definition

Options:

        C - Create new file
        A - Add sample data to old file
        D - Delete sample from file
        E - Edit sample data
        T - Type file on terminal
        P - Print file on printer
        X - Exit

Option...T

Pos  Job  Err  InJ  Tot  Col  Flow  DATA  RI UV IR VI  Opr      Sample ID
     No        No   InJ  Set  ml/m  pt/m
 1  6797  0    1    1    3    1.0   30    1  0  0  0   AFK  5872, PHOSPHATE MON

OK
>DIA 15C

Instrument No. 34 - HPGPC Sample Definition


Initials... AFK
Column Set... 3

Default Values:

Detectors (RI, UV, IR, VI)<RI>...
Flow Rate (ml/min)... 1
Data Collection Rate (pts/min)<60>... 30
Number of samples (1 - 16)<1>... 2

Position   1    Job No.   6803
Sample ID... TEST SAMPLE # 1
Customer... KUO

Position   2    Job No.   6804
Sample ID... TEST SAMPLE # 2
Customer... KUO

OK
>
```

Figure 1. Sample definition dialog for automated instrument operation.


```
DIA 16

INSTRUMENT NO. 34 - HPGPC

STARTING HPGPC SAMPLE ANALYSIS

OK
>
```

Figure 2. Dialog for initiation of automated instrument operation.

```
DIA 17

Instrument No. 34 - HPGPC Column Set Definition

Options:

        A - Add new column set
        D - Delete column set from file
        E - Edit column set data
        P - Print file on printer
        X - Exit

Option...A

DEFINE NEW HPGPC COLUMN SET

Column Set No.    4

Void Volume (ml)... 3
Total Volume (ml)... 8
Description....TEST COLUMN SET

OK
>DIA 17E

Instrument No. 34 - HPGPC Column Set Definition

Edit GPC Column Set

Column set... 4
Void volume (ml) <  3.0>... 3.2
Total volume (ml) <  8.0>...
Description <TEST COLUMN SET>
...

OK
>DIA 17D

Instrument No. 34 - HPGPC Column Set Definition

Delete GPC Column Set From File

Column set... 4

Column Set    4
TEST COLUMN SET

Delete this column set? Y

OK
>
```

Figure 3. Dialog for column set definition.

```
>DIA 18

Instrument No. 34 - HPGPC Calibration Curve Definition

Options:

        A - Add new calibration curve
        D - Delete calibration curve from file
        E - Edit calibration curve data
        P - Print file on printer
        X - Exit

Option...A

DEFINE NEW HPGPC CALIBRATION CURVE

Calibration Curve No.   12

Column Set... 1
Detector (RI, UV, IR, VI)... IR
Flow rate (ml/min)... 1
Solvent... THF
Temperature... 40
Polymer... POLYSTYRENE

Coefficients for Chain Length Calibration Curve:

        0
        1
        2
        3
        4
        5
        6
        7

Coefficients for Molecular Weight Calibration Curve:

        0   42.46
        1   -1.827
        2   .0356
        3   -.00183
        4
        5
        6
        7

OK
>DIA 18D

Instrument No. 34 - HPGPC Calibration Curve Definition

Delete HPGPC Calibration Curve

Curve No... 12

Curve  12  Column Set   1  Detector 3  22-OCT-82

Delete this calibration curve? Y

OK
>
```

Figure 4. Dialog for definition of calibration curves.

combines this information with the raw data and sample definition
information in memory.

The third stage is data transmission during which the
microcomputer transmits the entire data set for the sample to the
minicomputer. The data is stored on disk until the operator
initiates the fourth stage, data reduction. If multiple samples
and injections have been programmed, the minicomputer sends to
the microcomputer the information in the sample definition file
for the next sample, and operation continues without further
operator intervention.

Calibration is performed by using narrow molecular weight
distribution polystyrene standards. A polynomial up to sixth
order is fit to the $\log_{10}$(molecular weight) vs retention volume
data for the standards using conventional polynomial regression
methods, and the coefficients of the best fit polynomial (usually
fourth order or less) are used to define the calibration curve.
Dialogs 17 and 18, shown in Figures 3 and 4, are used by the
operator to define column sets and calibration curves. This
information is stored in files on the minicomputer until modified
or deleted by the operator and is used by the data analysis
programs.

An example of operator interaction with the primary analysis
program, GPC, is shown in Figure 5. The job number assigned by
the computer during sample definition is entered along with the
detector selected for analysis. The operator then selects the
baseline and the limits for data analysis by entering the times
of the desired points. The plots desired and the disposition of
the report file are chosen. The most recent calibration curve on
file for the column set is used by default but others may be
selected at the operator's option.

Integration of the data for the calculation of molecular
weight distribution averages is performed in time-volume space
using Simpson's Rule (assuming constant flow rate). Molecular
weight averages are calculated using the equation

$$\overline{M}_j = \frac{\int_{M_H}^{M_L} M^{j-1}(V)F(V)dV}{\int_{M_H}^{M_L} M^{j-2}(V)F(V)dV} \tag{1}$$

where $j = 1, 2, 3,$ and 4 correspond to the N, W, Z and Z+1
averages, respectively; M(V) represents the molecular weight
calibration curve as a function of retention volume and F(V) is
the normalized chromatogram height as a function of retention
volume. The weight differential molecular weight distribution,
$f_w(\log_{10}M)$, is calculated according to the method of Pickett et
al. (5) using the equation

$$f_w(\log_{10}M) = F(V) \cdot \frac{1}{2.303 \ (d\log_{10}M(V)/dV)} \qquad (2)$$

where $d\log_{10}M(V)/dV$ is the slope of the molecular weight
calibration curve. An example of the weight differential
molecular weight distribution plot is shown in Figure 8 along
with the weight cumulative molecular weight distribution. The
position of the molecular weight averages, $\bar{M}_n$, $\bar{M}_w$, $\bar{M}_z$, $\bar{M}_{z+1}$ and
$\bar{M}_{z+2}$ on the $\log_{10}M$ axis also are indicated in Figure 8. Examples
of the plots and report generated by the program are shown in
Figures 6-9. The report shown in Figure 9 is composed of four
sections: molecular weight distribution statistics, sample
information, raw chromatogram statistics and column set
information. The variance, skewness and kurtosis statistics
which involve moments about the mean are calculated from
equations relating moments about the mean to moments about the
origin (6). Customized plot presentations and coplotting of data
from multiple samples can be generated when required.

Other analysis methods dependent on multiple detectors
can be implemented using this automated system. Two methods
under development are the use of a continuous viscometer detector
with a refractive index detector to yield absolute molecular
weight and branching, utilizing the universal calibration curve
concept (4), and the use of a UV or IR detector with the
refractive index detector to measure compositional distribution
as a function of molecular weight.

Oligomer analysis is performed by a separate program OLIG by
a method analogous to conventional liquid chromatograph peak
analysis. This program utilizes the Digital Equipment
Corporation scientific subroutine PEAK. Since the subroutine
operates on progressively broadening peaks, the data is analyzed
in reverse order, that is, from long to short retention times.
The operator can select a baseline, or let the program select and
adjust the baseline automatically. Response factors may be
calculated at the operator's discretion, or concentrations can be
calculated from known response factors on an area basis. The
operator interaction with OLIG and samples of the report and plot
from this program are shown in Figures 10-12.

Conclusions

Benefits have been realized from the automation of the Waters
Model 150C ALC/GPC in several areas. First, a significant amount
of time has been saved by performing automated data collection
with automatic injection during night operation while unattended.
Secondly, record keeping is more complete and accurate. This has

```
>RUN $GPC

    JOB NUMBER, RUN NUMBER <1>                  9034
    SAMPLE 9165: 1306-12-B PMMA
    Detector: RI, UV, or IR  <RI>               RI
    CURVE 10,  13-AUG-82  - CHANGE ?            N
    PLOTS: RAW, VOLUME, MOLWT, CHAIN            RVM
    BASELINE IN MINUTES
                    START  <  23.93 >          36
                    STOP   <  79.87 >          75
    DATA LIMITS IN MINUTES
                    START  <  36.00 >          40
                    STOP   <  75.00 >          63
    OUTPUT FILE  (PRINT, SAVE) < DELETE >      P
            FILE HPGPC.LST CREATED
            PLOT FILES PRODUCED:
                    RAWDAT
                    RVOL
                    MOLWT
```

Figure 5.   Operator interaction with program GPC.

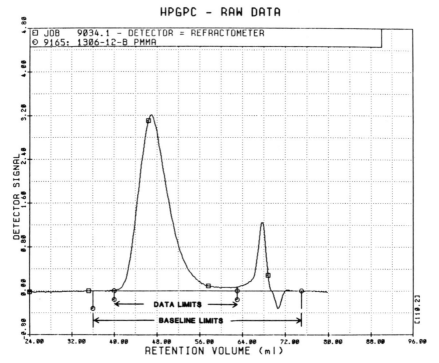

Figure 6.   Raw HPGPC data with operator selected baseline.

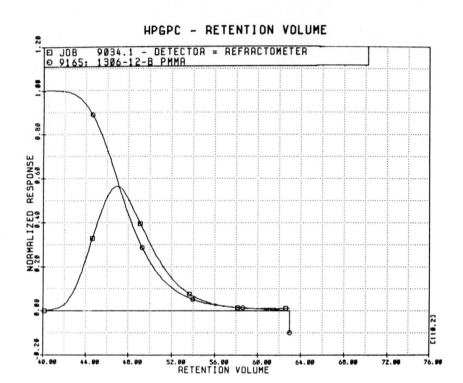

Figure 7.   Baseline corrected normalized retention volume data over the region to be analyzed.

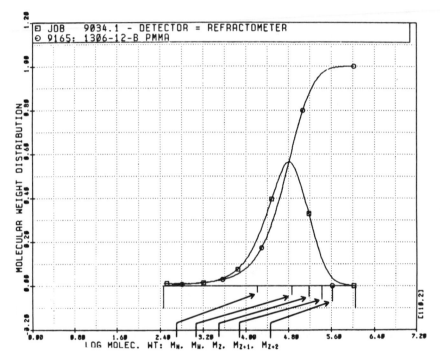

Figure 8. Weight differential and integral molecular
weight distribution plots.

```
JOB   9034.1   HPGPC - REFRACTIVE INDEX DETECTOR ------------------------

      9165: 1306-12-B PMMA

MOLECULAR WEIGHT DISTRIBUTION -----------------------------------------------
                                 MEAN      VARIANCE     SKEWNESS    KURTOSIS
      NUMBER AVERAGE          0.163E+05   0.929E+09   0.461E+01   0.409E+02
      WEIGHT AVERAGE          0.732E+05   0.568E+10   0.302E+01   0.166E+02
      Z AVERAGE               0.151E+06   0.173E+11   0.237E+01   0.819E+01
      Z+1 AVERAGE             0.266E+06
      Z+2 AVERAGE             0.416E+06
      LOW MOLECULAR WEIGHT                            0.291E+03
      HIGH MOLECULAR WEIGHT                           0.110E+07
      POLYDISPERSITY, MEAN WT/MEAN NUMBER               4.4906
      PEAK MOLECULAR WEIGHT                           0.633E+05
      SLOPE OF CALIBRATION CURVE 10 AT PEAK            -0.153389

**   NO COEFFICIENTS FOR CHAIN LENGTH IN CALIBRATION CURVE 10

SAMPLE INFORMATION _____    _____

      REQUESTED BY GARY CARLSON          SAMPLE VOLUME, UL            200
      INSTRUMENT NUMBER         34       BASELINE TIME, LOW         36.00
      OPERATOR                 AFK       BASELINE TIME, HIGH        75.00
      RUN DATE           11-OCT-82       BASELINE SLOPE         -0.000952
      RUN TIME            22:53:37       DATA TIME, LOW             40.00
      SENSITIVITY            0032        DATA TIME, HIGH            63.00
      SCALE FACTOR            32         NUMBER OF DATA POINTS        690

RAW CHROMATOGRAM STATISTICS -----------------------------------------------

      MEAN VOLUME            48.08       PEAK TIME                  47.00
      VARIANCE              10.6909      PEAK VOLUME                47.00
      SKEWNESS              1.1792       PEAK HEIGHT               3.2058
      KURTOSIS              2.4579       MOMENT 3 ABOUT MEAN      41.2097
      AREA                 21.2442       MOMENT 4 ABOUT MEAN     625.1477

COLUMN SET -------------------------------------------------------------
      Waters u-Styragel, 6-col, 1000000-100000-10000-1000-500-100

      COLUMN SET NUMBER          3       CALIBRATION -----------------------
      FLOW RATE, ML/MIN        1.0
      COLUMN TEMPERATURE       50C       CURVE NUMBER                  10
      SOLVENT                  THF       CALIB. DATE            13-AUG-82
      VOID VOLUME             34.0       CALIB. TEMP.               50.0
      TOTAL VOLUME            75.0       CALIB. POLYMER:     PS 4th order
```

Figure 9.   Sample report from program GPC.

```
RUN OLIG

Job and run numbers ?      10758,1
9344: RESIMENE

Enter editing limits in minutes ( ^Z for full range of data )
    Lower limit    < 10.52 >    21
    Upper limit    < 44.97 >    31
Enter baseline limits in minutes ( ^Z for computed baseline )
    Lower limit    < 21.00 >    20
    Upper limit    < 31.00 >    37
Dens, Base, Gate, Diff, Reset  < 3, 8, 2, 2, R >      1,18,,,N

9344: RESIMENE
```

#	PEAK TIME	% AREA	---TIME--- START	END	-----HEIGHT----- START	PEAK	END	HALF-WIDTH	AREA	PT DENS
1	25.33	20.43	21.73	25.53	98.	251.	248.	0.200	261.47	2.
2	26.10	13.67	25.53	26.57	248.	298.	234.	0.367	174.93	2.
3	27.27	21.86	26.57	28.07	234.	405.	160.	0.467	279.74	2.
4	29.45	44.04	28.07	30.95	160.	648.	106.	0.583	563.65	2.

```
                                                          S

Print the file ? (No/Yes/Plot/Edit/Save/Response factors)
Files created: OLIG.T1,T2 and OLIG.LST

^
```

Figure 10. Operator interaction with program OLIG.

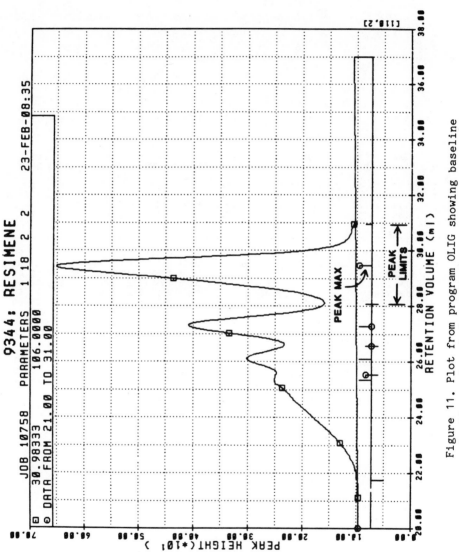

Figure 11. Plot from program OLIG showing baseline selection and peak locations.

```
JOB  10758.]   HPGPC - REFRACTIVE INDEX DETECTOR --------------09-MAR-09:49

     9344: RESIMENE

    PEAK    %   ----TIME----    ------HEIGHT-----   HALF-            PT
 #  TIME   AREA START   END     START  PEAK   END   WIDTH    AREA   DENS

 1  25.33  20.43 21.73  25.53      98.  251.  248.  0.200  261.47   2.
 2  26.10  13.67 25.53  26.57     248.  298.  234.  0.367  174.93   2.
 3  27.27  21.86 26.57  28.07     234.  405.  160.  0.467  279.74   2.
 4  29.45  44.04 28.07  30.95     160.  648.  106.  0.583  563.65   2.

SAMPLE INFORMATION _____

     REQUESTED BY SWAFFORD              SAMPLE VOLUME, UL              100
     INSTRUMENT NUMBER         34       DATA TIME, LOW              21.00
     OPERATOR                 AFK       DATA TIME, HIGH             31.00
     RUN DATE            22-FEB-83      BASELINE TIME, LOW          20.00
     RUN TIME            13:04:47       BASELINE TIME, HIGH         37.00
     NUMBER OF DATA POINTS    600       BASELINE SLOPE           0.625000

COLUMN SET --------------------------------------------------------------

     VARIAN MICROPAK TSK - 2000 H, 3000 H
                                  TUNING PARAMETERS ------------------
     COLUMN SET NUMBER          1       ORIGINAL POINT DENSITY         1
     FLOW RATE, ML/MIN        1.0       BASELINE TEST                 18
     VOID VOLUME             15.0       GATE FACTOR                    2
     TOTAL VOLUME            35.0       MINIMUM DIFFERENCE             2
     COLUMN TEMPERATURE       50C       POINT DENSITY NOT RESET
```

Figure 12. Sample report from program OLIG.

simplified accurate reproduction of experimental results and has helped discern subtle or long term variability in the operating characteristics of the instrument. Finally it has facilitated the development of experimental methodology for non-standard detectors.

Literature Cited

1. Niemann, T. F.; Koehler, M. E.; Provder, T. "Microcomputers used as Laboratory Instrument Controllers and Intelligent Interfaces to a Minicomputer Timesharing System" in "Personal Computers in Chemistry"; Lykos, P., Ed.; John Wiley and Sons: New York, 1981; pp. 85-91.
2. Kah, A. F.; Koehler, M. E. ; Niemann, T. F.; Provder, T.; Eley, R. R. "An Automated Ferranti-Shirley Viscometer" in "Computer Applications in Applied Polymer Science"; Provder, T., Ed.; ACS SYMPOSIUM SERIES No. 197, American Chemical Society: Washington, D.C., 1982; pp.223-241.
3. Kah, A. F.; Koehler, M. E.; Grentzer, T. H.; Niemann, T. F.; Provder, T. "An Automated Thermal Analysis System for Reaction Kinetics" in "Computer Applications in Applied Polymer Science"; Provder, T., Ed.; ACS SYMPOSIUM SERIES No. 197, American Chemical Society: Washington, D.C., 1982; pp. 197-311.
4. Malihi, F. B.; Kuo, C.; Koehler, M. E.; Provder, T.; Kah, A. F. "Development and Application of a Continuous GPC Viscosity Detector for the Characterization of Absolute Molecular Weight Distribution of Polymers", this volume.
5. Pickett, H. E.; Cantow, M. J. R.; Johnson, J. F. Appl. Polym. Sci. 1966, 10, 917.
6. Aitken, A. C. "Statistical Mathematics"; Oliver and Boyd: London, 1962; Chap. 2.

RECEIVED October 4, 1983

Comparison of Size Exclusion Chromatography Calibration Techniques

Using Narrow and Broad Molecular Weight Distribution Standards

THOMAS V. ALFREDSON, LORI TALLMAN, and WILLIAM J. PERRY

Varian Instrument Group, Walnut Creek, CA 94598

Size exclusion chromatography (SEC) polymer elution profiles yield information regarding the molecular size distributions of polydisperse macromolecules. Polymer molecular weight distribution (MWD) represents an intrinsic property which provides direct correlation with many end-use physical properties and a universal criterion for polymer characterization (1). In order to convert elution profiles or chromatograms into MWD information proper calibration methods are required. SEC molecular weight calibration techniques represent experimental approaches for transformation of polymer elution profiles into MWD information and are dependent upon instrumentation, columns, and the polymer/solvent system under study.

SEC calibration methods can be generally categorized into techniques which employ a series of narrow MWD standards and those which employ one (or more) broad MWD standards (2). Calibration techniques which utilize polydisperse, broad MWD standards have been found to be particularly useful when narrow MWD standards are not available or universal calibration methodology is impractical as for example with most water-soluble polymers or polymer/solvent/temperature combinations for which appropriate Mark-Houwink constants are not known or readily available.

Methods of molecular weight calibration using polydisperse standards are of two fundamental types. Techniques which utilize a polydisperse standard with known molecular weight distribution (referred to as integral methods) and those which make use of one or more broad MWD standards for which any pair of $\bar{M}_n$, $\bar{M}_w$ or $\bar{M}_v$ values are known and assume a linear molecular weight calibration curve (referred to as linear methods). Methodologies have also been developed for SEC calibration which employ a polydisperse standard and use the universal molecular weight calibration curve obtained with a series of narrow MWD polystyrene standards (3) or use a polydisperse standard to calculate effective Mark-Houwink constants for utilization of universal calibration approaches (4).

Integral methods of SEC calibration which make use of a well

0097-6156/84/0245-0073$06.75/0

characterized, broad MWD standard, such as the method developed
by Cantow et al. (5), correlate molecular weights and elution
volumes by successively super-imposing the cumulative molecular
weight distribution and the integrated, normalized SEC chromatogram. Historically, the well-characterized molecular weight
distribution of a polydisperse standard was experimentally
obtained by column fractionation in which the molecular weight
of each fraction was determined by conventional methods such as
light scattering and osmometry. Weis and Cohn-Ginsberg (6)
developed an alternative procedure based upon theoretical
polymer molecular weight distribution shape and known average
molecular weight values in order to yield the required molecular
weight distribution information necessary for use of the polydisperse standard in calibration. Characterization of organic-
soluble polymers by Swartz et al. (7) and water-soluble polymers
by Abdel-Alim and Hamielec (8) has been accomplished using this
approach to SEC calibration. In general, the information
required of a polydisperse standard by integral calibration
methods is not readily available or can be time consuming to
generate. Prediction of molecular weight distribution shape also
is not always straightforward. Thus the utility of integral
methods using polydisperse standards for calibration has been
limited compared to linear methods using polydisperse standards.

Linear calibration methods employing a broad MWD standard
with a known pair of $\bar{M}_n$, $\bar{M}_w$ or $\bar{M}_v$ values and assume a linear
molecular weight calibration curve are based upon a method
originally developed by Frank et al.(9) which relied upon known
$\bar{M}_n$ and $\bar{M}_w$ values of a single broad MWD standard and utilized a
graphical approximation method to obtain a working calibration
curve. Balke, Hamielec, LeClair and Pearce (10) developed a
much improved refinement to this technique by replacing the
graphical approximation method with a computer program using a
Rosenbrock search routine to determine a linear calibration
curve. Such an iterative, two variable search routine was
employed to develop a calibration curve which can be expressed
as follows:

$$V_e = C_1 - C_2 \log_{10} (M) \qquad\qquad (1)$$

where

V_e = elution volume
M = molecular weight
C_1 and C_2 = constants to be found by computer search
program.

Using a search routine, a computer program can calculate the
constants C_1 and C_2 from definitions of the moments of the distribution:

$$\bar{M}_w = \sum_i W_i \, M_i \tag{2}$$

$$\bar{M}_n = \frac{1}{\sum_i (W_i/M_i)} \tag{3}$$

where

W_i = weight fraction
M_i = molecular weight

Substituting the expression for M listed in eqn (1) into eqns (2) and (3) it is easily shown that:

$$\bar{M}_w = \sum_i W_i \, 10^{(C_1 - V_i/C_2)} \tag{4}$$

$$\bar{M}_n = \frac{1}{\sum_i W_i/10^{(C_1 - V_i/C_2)}} \tag{5}$$

The Rosenbrock search routine in the computer program employed by Balke and Hamielec was found to converge to the correct optimum values of C_1 and C_2 within approximately 200 iterations.

Loy (11) has published a procedure based upon the method of Balke and Hamielec which uses a more efficient iterative, single-variable search algorithm which relies upon the fact that the dispersity ($\bar{M}_w/\bar{M}_n$) is a function of C_2 only. The computer program incorporating this much faster algorithm converges to the optimum C_2 within 36 iterations.

Pollock et al. (12) have also exploited the fact that poly dispersity index is a function of C_2 only in a study utilizing a Monte-Carlo simulation technique to compare error propagation in the method of Balke and Hamielec to a revised method (GPCV2) proposed by Yau et al. (13) which incorporated correction for axial dispersion.

Malawer and Montana (14) have developed an efficient iterative, sequential single variable search algorithm which relies upon a polydisperse standard with known $\bar{M}_n$ and $\bar{M}_v$ values. A direct graphical proof of the algorithm was presented.

A reliable and very rapid search algorithm in the computer program for use in linear calibration methods is highly desirable when data processing is performed with a microcomputer for automated, on-line calibration and MWD calculations. A fast search algorithm requiring few iterations ensures that convergence will be achieved within a few minutes even on an inexpensive, small personal computer. With this goal in mind, a proprietary iterative, two-variable search algorithm has recently been developed in our laboratory and incorporated into a user-interactive computer program for SEC polymer characterization (15). The highly efficient search algorithm is used in a linear calibration method based upon that of Balke and Hamielec which employs a single, broad MWD standard with known $\bar{M}_n$ and $\bar{M}_w$ values. The calibration curve in this method can be described as follows:

$$\log 10 \ (M) \ = \ C_o + C_1 \ V_e \qquad\qquad (6)$$

where

    M = molecular weight
    V_e = elution volume
    C_o, C_1 = constants to be found by search algorithm

The search algorithm employs a successive approximation and accelerated convergence technique on the independent variable in eqn (6), then approximates the dependent variable from the simultaneous solution of the equations for $\bar{M}_n$ and $\bar{M}_w$ moments of the polydisperse standard distribution. Convergence to within 0.1% of true $\bar{M}_n$ and $\bar{M}_w$ values of a broad MWD standard is usually achieved in six to nine iterations.

SEC calibration methods which employ a series of narrow MWD standards are based upon a peak position method and traditionally have been the most widely practiced calibration procedures. The peak position method simply correlates the peak elution volume of each standard to its nominal molecular weight or size value. A curve fitting procedure (usually a least squares regression) is used to obtain a working calibration curve. The serious limitation of polymer chemical types for which a series of narrow MWD standards covering a wide molecular weight range can be obtained led to the development of experimental approaches which could be applied to polymer chemical types other than that of the narrow MWD standards employed in calibration.

The Q-factor approximation method (16) was an early attempt at extending the application of the peak position method to polymers of different chemical types than the calibration standards.

The Q-factor approach is based upon the weight-to-size ratios (Q-factors) of the calibration standard and the polymer to be analyzed. The Q-factors are employed to transform the calibration curve for the chemical type of the standards (e.g. polystyrene) into a calibration curve for the chemical type of polymer under study. The inherent assumption in such a calibration approach is that the weight-to-size ratio is not a function of molecular weight but a constant. The assumption is valid for some polymer types (e.g. polyvinylchloride) but not for many polymer types. Hence the Q-factor method is generally referred to as an approximation technique.

A direct consequence of the development of hydrodynamic volume theory in SEC has been the universal calibration method as introduced by Benoit (17). Universal calibration methodology is based upon the fact that retention in SEC can be described as a function of the hydrodynamic volume of polymer molecules.

Usually the function $[(\eta) \cdot M]$ (intrinsic viscosity times molecular weight) is used to represent hydrodynamic volume which is
plotted versus elution volume. For such a plot the calibration
curves of many polymers fall on the same line irrespective of
polymer chemical type. Universal calibration methodology usually
requires knowledge of Mark-Houwink constants for the polymer/
temperature/solvent system under study.

The purpose of this study was to evaluate the linear calibration technique employing a single polydisperse standard and
the search algorithm described above for non-aqueous and aqueous
SEC. Comparison of this calibration technique to peak position,
universal calibration, and Q-factor approximation techniques
which make use of a series of narrow MWD polystyrene standards
was also carried out.

Experimental Techniques. Chromatography was performed on a
Varian model 5060 HPLC equipped with a RI-3 refractive index
detector. A Vista Plus Gel Permeation Chromatography (GPC) data
system was used consisting of a Vista 401 chromatography data
system serially connected to an Apple II microcomputer. The
Vista 401 performs data acquisition and allows data storage and
automations capability while all SEC data processing is performed
on the Apple II by means of user-interactive GPC software for
automated, on-line calibration and polymer analysis.

Non-Aqueous SEC Experiments. Non-aqueous SEC separations were
carried out at ambient temperatures using two Varian MicroPak
TSK GMH6 columns in series (7.5mm i.d. x 30cm each). This
column is a mixed bed column containing pore sizes from 250 Å to
10^7 Å blended to ensure linearity of the molecular weight calibration curve. The mobile phase employed tetrahydrofuran at a
flow rate of 1 ml/min. Sample injection volumes were 50μℓ
using a Rheodyne 7126 manual loop injector.

Samples of narrow MWD polystyrene standards were obtained
from Toyo Soda Mfg. Co., Ltd. (Tokyo, Japan) of the following
molecular weights:

Designation	MW	$(\bar{M}_w/\bar{M}_n)$	Designation	MW	$(\bar{M}_w/\bar{M}_n)$
A-500	5×10^2	1.15	F-10	1.07×10^5	1.01
A-1000	1×10^3	1.15	F-20	1.86×10^5	1.07
A-2500	2.8×10^3	1.05	F-40	4.22×10^5	1.05
A-5000	6.2×10^3	1.04	F-80	7.75×10^5	1.01
F-1	1.02×10^4	1.02	F-126	1.26×10^6	1.05
F-2	1.67×10^4	1.02	F-240	2.42×10^6	1.09
F-4	4.28×10^4	1.01			

A polydisperse polystyrene standard was obtained from Dow
Chemical Co. (Midland, Michigan). This material was designated
Dow 1683 polystyrene standard and has been well characterized
with reported values as follows (18):

$\bar{M}_n$ = 100,000
$\bar{M}_w$ = 250,000
Polydispersity Index = 2.5

Dow 1683 polystyrene standard was utilized as a broad MWD
standard in the linear calibration method due to its distribution
symmetry and particular lack of significant tail at the low end
of its MWD.

NBS 706 broad distribution polystyrene, NBS 705 narrow distribution polystyrene, and NBS 1478 narrow distribution polystyrene reference materials were used as samples in the evaluation of the proposed linear calibration method in this study.
These reference materials have the following reported values:

NBS Standard Reference Material	$\bar{M}_w$**	$\bar{M}_n$	($\bar{M}_w/\bar{M}_n$)	Concentration injected (%W/V)
NBS 706 broad polystyrene	257,800	122,700	2.1	0.15%
NBS 705 narrow polystyrene	179,300	170,900	1.05	0.15%
NBS 1478 narrow polystyrene	37,400	35,800	1.05	0.20%

** Measured by light scattering

Polystyrene Mark-Houwink constants of K=1.6×10^{-4} dg/ℓ and
a = 0.706 were employed (19).

Polyvinyl chloride (PVC) broad MWD standard obtained from
Polysciences Inc. (Warrington, Penn.) was employed in the linear
calibration method in a study of calculated molecular weight
accuracy as a function of calibration methodology. The PVC polydisperse standard (lot #5-0069) had reported values of $\bar{M}_w$ = 83,500
and $\bar{M}_n$ = 37,100. In this study the accuracy of results using the
linear calibration method was compared to the accuracy of results
using a Q-factor approximation method and universal calibration
methodology employing a series of narrow MWD polystyrene standards. A PVC polymer with $\bar{M}_w$ = 152,000 (measured by light scattering) was used as the sample in the study of accuracy of calculated molecular weight as a function of calibration method.
A concentration of 0.15% was injected for each PVC material. PVC
Mark-Houwink constants of K=1.63×10^{-4} dg/ℓ and a=0.766 were
employed (20).

Aqueous SEC Experiments. Aqueous SEC separations were carried out
at ambient temperature using two column sets of MicroPak TSK
PW Type gel which were investigated for linearity of molecular
weight calibration curve using polyethylene glycol (PEG) and
polyethylene oxide (PEO) narrow MWD standards. Columns were
matched in pore volume as closely as possible to promote linearity
of the molecular weight calibration curve. Column set A consisted

of MicroPak TSK 3000PW + 4000PW + 5000PW + 6000PW (7.5mm i.d. x 30cm each) columns in series. Column set B consisted of MicroPak 3000PW + 3000PW + 6000PW + 6000PW (7.5mm i.d. x 30cm each) columns in series. Mobile phase for analysis of PEG and PEO standards was 50mM sodium sulfate at a flow rate of 1 ml/min. Injection volume was 100µℓ.

Samples of narrow MWD PEG standards were obtained from Fluka Chemical Co., (Hauppauge, NY) of the following molecular weights:

Designation	MW	Designation	MW
PEG 400	$4 \times 10^2 \pm 5\%$	PEG 4000	$4 \times 10^3 \pm 12\%$
PEG 600	$6 \times 10^2 \pm 5\%$	PEG 6000	$6 \times 10^3 \pm 12\%$
PEG 1540	$1.54 \times 10^3 \pm 9\%$	PEG 10K	$1 \times 10^4 \pm 15\%$
PEG 2000	$2 \times 10^3 \pm 8\%$		

Samples of narrow MWD PEO standards were obtained from Toyo Soda Mfg. Co. Ltd (Tokyo, Japan) of the following molecular weights:

Designation	MW	$(\bar{M}_w/\bar{M}_n)$	Designation	MW	$(\bar{M}_w/\bar{M}_n)$
SE-2	2.5×10^4	1.14	SE-30	2.8×10^5	1.05
SE-5	4.0×10^4	1.03	SE-70	6.6×10^5	1.10
SE-8	7.3×10^4	1.02	SE-150	1.2×10^6	1.12
SE-15	1.5×10^5	1.04			

Sample concentrations were 0.3% w/v and 0.15% w/v respectively for the PEG and PEO standards. To aid dissolution of the PEO standards, 0.5% ethanol was added to the aqueous solutions.

Dextrans were employed to evaluate the linear calibration method for utility in aqueous SEC. The dextrans were obtained from Pharmacia Chemical Co. (Upsala, Sweden) of the following molecular weights:

Designation	$\bar{M}_w$	$\bar{M}_n$	$(\bar{M}_w/\bar{M}_n)$	Comments
Dextran T-70	64,200	44,000	1.46	—
Dextran T-40	39,900	26,200	1.52	T-40 used as sample for evaluation
Dextran Blend	52,050	32,850	1.58	Blend (1 part T-70 and 1 part T-40) used as polydisperse calibration standard.

The blend of T-70 and T-40 dextran materials was utilized as a polydisperse calibration standard for the linear calibration method and the T-40 dextran standard was used as a sample for evaluation. Concentrations of 0.15% W/V were injected for each dextran material chromatographed.

Results and Discussion

Non-Aqueous SEC Evaluation. The SEC calibration report for a
peak position method using a series of narrow MWD polystyrene
standards is shown in Table I, As can be seen, a linear fit produces a high correlation to the data (r=0.9997). Figure 1 displays the molecular weight calibration plot of elution volume
versus log molecular weight for the series of polystyrene standards.

Table I.   Calibration Report for a Series of Narrow MWD PS
                          Standards

	Least-Squares Curve Fit To Data				
Curve Type	Correlation Coefficient	A	B	C	D
Y=A+B(X)	.9997	11.1495	-.4046		

$Y = $ Log (Mol. Weight)
$X = $ Elution Volume (mls)

Actual Elution Vol.	Actual Log (Mol. Wt.)	Actual Mol. Wt.	Calculated Log (Mol. Wt.)	Calculated Mol. Wt.	% Difference
11.7	6.4609	2890000	6.4156	2604020	-9.9
12.46	6.1004	1260000	6.1081	1282768	1.81
12.92	5.8893	775000	5.922	835665	7.83
13.6	5.6253	422000	5.6469	443511	5.1
14.51	5.2695	186000	5.2787	189984	2.14
15.1	5.0204	107000	5.04	109649	2.48
16.17	4.6314	42800	4.6071	40465	5.46
17.15	4.2227	16700	4.2106	16240	-2.76
17.74	4.0086	10200	3.9719	9373	-8.11
18.13	3.7924	6200	3.8141	6517	5.12
19	3.4472	2800	3.4621	2898	3.49

     Table II shows the report from the linear calibration method
employing the Dow 1683 broad MWD polystyrene standard. As can be
seen in the report, the elution volume profile of the polydisperse
standard contained 122 area/time slices upon which calibration
calculations were based. The correlation coefficient listed in
this report is an index of degree of fit of the calibration curve
based upon the difference between calculated and true values of
molecular weight averages of the standard. A plot of the linear
calibration method molecular weight calibration curve is displayed
in Figure 2. A comparison of molecular weight calibration curve
plots for calibration using a peak position method with a series
of narrow MWD standards (Figure 1) and a linear calibration method
with a polydisperse standard (Figure 2) reveals that both calibration curves cover approximately the same elution volumes. This is
not surprising due to the fact that both methodologies use standards which cover similar molecular weight ranges - about 3000 to
2×10^6. As can be seen by comparing Table I and II, the curve
coefficients are very similar with almost identical slopes.

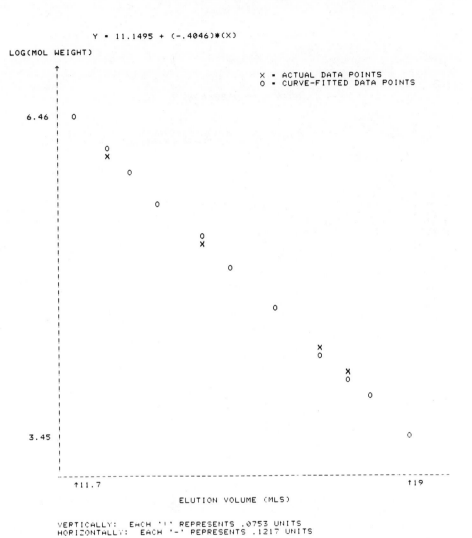

Figure 1. Calibration curve for a series of narrow MWD
PS standards. Linear least squares fit for log (MW) versus
elution volume.

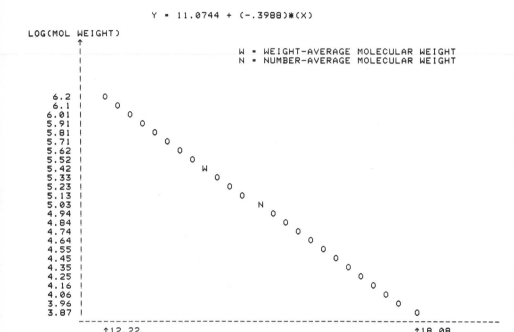

HORIZONTALLY: EACH '-' REPRESENTS .244 UNITS

Figure 2.   Calibration curve for polydisperse PS standard
using a linear method.   Plot of log (MW) versus elution
volume.

Table II. Calibration Report for Polydisperse PS Standard

Curve Type	Correlation Coefficient	A	B	C	D
Y=A+B(X)	1.0005	11.0744	-.3988		

Y = Log (Mol. weight)
X = Elution Volume (mls)

	Calculated	Actual
Number - average molecular weight:	100091	100000
Weight - average molecular weight:	250227	250000
Number of slices: 122 Slice width:	3 seconds/slice	

NBS 706 broad polystyrene, NBS 705 narrow polystyrene, and NBS 1478 polystyrene reference materials were used as samples for investigation of molecular weight accuracy as a function of calibration method. A peak position method using a series of narrow MWD polystyrene standards was compared to a linear calibration method utilizing a single broad MWD standard (Dow 1683 polystyrene standard). SEC peak processing parameters used for calculation of MWD values were held constant. Molecular weight averages were calculated for each NBS reference material by means of each calibration method. Table III lists the results of this study. Overall no significant difference in accuracy of calculated $\bar{M}_w$ and $\bar{M}_n$ values can be discerned on the basis of calibration methodology.

Table III. Molecular Weight Accuracy as a Function of Calibration Technique

	$\bar{M}_w$	% Difference	$\bar{M}_n$	% Difference
NBS 705 Narrow PS Reference STD				
Reported value	179,300	—	179,900	—
Peak Position Method	173,509	- 3.2%	146,174	- 14.5%
Linear Method Using Polydisperse Std.	177,433	- 1.0%	150,246	- 12.1%
NBS 706 Broad PS Reference STD				
Reported Value	257,800	—	122,700	—
Peak Position Method	276,055	+ 7.1%	123,345	+ 0.5%
Linear Method Using Polydisperse Std.	279,574	+ 8.4%	127,870	+ 4.2%
NBS 1478 PS Reference STD.				
Reported Value	37,400	—	35,800	—
Peak Position Method	35,716	- 4.5%	33,487	- 6.5%
Linear Method Using Polydisperse Std.	37,389	< 0.1%	35,121	- 1.9%

The NBS 706 broad polystyrene reference material is very
similar in molecular weight and dispersity to the Dow 1683 polystyrene standard used for calibration. The NBS 705 narrow polystyrene material is lower in molecular weight and much narrower
in dispersity than the Dow 1683 polystyrene standard and the NBS
1478 polystyrene reference material is significantly lower in
molecular weight and dispersity. However, all three reference
materials elute within the calibrated elution volume range of the
molecular weight calibration curve generated by the linear calibration method utilizing the Dow 1683 polystyrene standard. SEC
chromatograms of the Dow 1683 polystyrene standard and the NBS
polystyrene reference materials are shown in Figure 3.

In a similar study by Yau et al. (21) which compared molecular weight accuracy as a function of a peak position calibration
method and a linear polydisperse standard method using polystyrene standards, it was found that the peak position method using
a series of narrow MWD standards gave more accurate results for
narrow polydispersity samples and the linear calibration method
gave more accurate results on samples of polydispersity similar
to the polydisperse standard used for calibration. Based upon
the results of the present study, it appears that accuracy is not
a function of polydispersity of the sample when a linear, polydisperse standard method is used for calibration provided that
the samples being analyzed elute within the calibrated elution
volume range of the polydisperse standard and axial dispersion
is minimized by use of high efficiency SEC columns. Axial
dispersion for the non-aqueous SEC experiments was measured
and found to be < 5% over the molecular weight range of interest. Since dispersion was found to be minimal, no corrections
to calculated molecular weight values were made.

Polyvinyl chloride polymers were utilized in a separate study
to evaluate calculated molecular weight accuracy as a function
of a universal calibration method, a Q-factor approximation
method, and a linear calibration method employing a polydisperse
standard. Table IV displays the calibration report generated
from a linear calibration method using a broad MWD PVC calibration standard. At a slice rate of 3 second/slice, 130 slices
were used in the calibration calculations to define the elution
volume profile of the polydisperse PVC standard. Table V lists
the calibration report obtained from use of a series of narrow
MWD polystyrene standards utilizing universal calibration methodology. A comparison of calibration curves generated with a series
of narrow MWD polystyrene standards utilized in a universal calibration method (Table V) and a linear polydisperse standard calibration method (Table IV) shows that the calibration curve
coefficients are very similar.

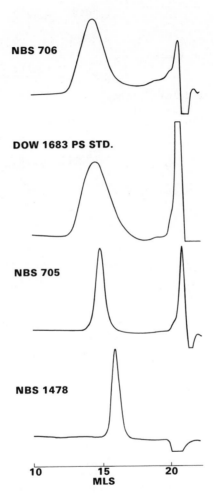

Figure 3.   Chromatograms of NBS 706, Dow 1683, NBS 705, and
NBS 1478 standard polystyrene reference materials.
Detector: Refractive Index.

Table IV. Calibration Report for Polydisperse PVC Standard

Curve Type	Correlation Coefficient	A	B	C	D
Y=A+B(X)	.9996	11.3079	-.4252		

Y = Log (Mol. Weight)
X = Elution volume (mls)

	Calculated	Actual
Number - Average Molecular Weight:	37068	37100
Weight - Average Molecular Weight:	83430	83500
Number of Slices: 130	Slice Width: 3 seconds/slice	

Table V. Universal Calibration Method Report for PVC

Least-Squares Curve Fit To Data

Curve Type	Correlation Coefficient	A	B	C	D
Y=A+B(X)	.9997	10.7639	-.39		

Y= Log (Mol. Weight)
X= Elution Volume (mls)

Actual Elution Vol.	Actual Log (Mol. Wt.)	Actual Mol. Wt.	Calculated Log (Mol. Wt.)	Calculated Mol. Wt.	% Difference
11.7	6.2441	1754284	6.2005	1586657	- 9.56
12.46	5.8966	788134	5.9041	801785	1.73
12.92	5.6931	493287	5.7246	530448	7.53
13.6	5.4386	274536	5.4594	288019	4.91
14.51	5.0956	124624	5.1045	127201	2.07
15.1	4.8641	73131	4.8744	74881	2.39
16.17	4.4805	30234	4.457	28644	- 5.26
17.15	4.0865	12204	4.0748	11880	- 2.66
17.74	3.8801	7588	3.8447	6993	- 7.84
18.13	3.6717	4696	3.6926	4927	4.92
19	3.3389	2182	3.3532	2256	3.37

   Table VI lists the results of this study of calculated
molecular weight accuracy as a function of calibration method with
the PVC polymers. A PVC polymer sample was analyzed and molecular
weight averages were calculated by means of each calibration
method. All SEC peak processing parameters used for calculation
of MWD values were held constant. As shown in Table VI, the
universal calibration method provided a somewhat more accurate M_w
value than the Q-factor approximation method or the linear, polydisperse standard method.

Table VI.   Molecular Weight Accuracy As A Function of Calibration
Technique for PVC Polymer Sample

	$\bar{M}_w$	% Difference	$\bar{M}_n$
Reported Value	$152,000^*$	—	—
Q-Factor Approx. Method**	136,093	-10.5%	58,718
Universal Calibration Method	146,876	- 3.4%	67,318
Linear Method Using Poly-disperse Std.	166,024	+ 9.2%	65,662

* Determined by light scattering
** $Q_{pvc} = 25$

Universal calibration methodology has been successfully
applied to PVC and many other chemical types of polymers (22).
However, availability of Mark-Houwink constants can limit the
utility of universal calibration. In these cases, the use of a
linear, polydisperse standard calibration method is a viable
alternative for generation of a molecular weight calibration
curve.

Aqueous SEC Evaluation. A comparison of SEC calibration reports
from peak position methods using a series of PEG and PEO narrow
MWD standards is shown in Table VII for the two MicroPak TSK Type

Table VII.   Comparison of Calibration Reports Using A Series of
Narrow MWD Polyethylene Oxide Standards for MicroPak
TSK PW Column Sets

Column Set A: 3000PW + 4000PW + 5000PW + 6000PW		Column Set B: 3000PW + 3000PW + 6000PW + 6000PW
Standard Type: Narrow Standards		Narrow Standards
Calibration Basis: Molecular Weight		Molecular Weight
Curve Type: $Y = A + B (X)$		$Y = A + B (X)$
Correlation Coefficient: 0.995		0.996
Curve Coefficients: $A = 10.9739$ $B = -.2135$		$A = 13.1186$ $B = -.267$

PW gel column sets investigated for linearity. A least squares
polynomial curvefitting procedure was employed to determine a
first order polynomial fit to the data. The column set consisting of MicroPak TSK 6000PW + 6000PW + 3000PW + 3000PW (7.5 mm
i.d. x 30 cm each) had a slightly higher correlation coefficient
(r = 0.996). Figure 4 displays the plots of elution volume versus log molecular weight for each column sets molecular weight
calibration curve. Although column set A (3000PW + 4000PW +
5000PW + 6000PW) provided slightly higher resolution (i.e. lower
slope), column set B (3000PW + 3000PW + 6000PW + 6000PW) was
chosen for subsequent experiments on the basis of higher correlation to a linear fit of the data.

Dextran polymers were used to evaluate the utility of the
linear, polydisperse calibration method for water-soluble polymer
characterization. A blend of T-40 and T-70 dextran standards was
used as a polydisperse calibration standard. Table VIII displays
the report from the linear calibration method using this standard.
Nine iterations of the search algorithm were required for converg-
ence to the true $\overline{M}_w$ and $\overline{M}_n$ values of the standard. As can be seen
in the report, the elution volume profile of the standard con-
tained 72 area/time slices upon which calibration calculations
were based. The slice width was set at 10 seconds/slice. Figure
5 shows a plot of the calibration curve generated from the linear
calibration method utilizing the dextran standard.

Table VIII. Calibration Report For Polydisperse Dextran Standard

Curve Type	Correlation Coefficient	A	B	C	D
Y = A+B(X)	.9995	9.0638	− .1399		
Y= Log (Mol. Weight)					
X= Elution Volume (mls)			Calculated		Actual
Number − Average Molecular Weight:			32805		32850
Weight − Average Molecular Weight:			51999		52050
Number of Slices: 72	Slice Width:10 seconds/slice				

A T-40 dextran standard material was used as a sample for
evaluation of calculated molecular weight accuracy using the
linear calibration method. Results of this study are shown in
Table IX. Errors of 5.3% in $\overline{M}_w$ and 14.8% in $\overline{M}_n$ were found.
Hamielec and Omorodion (23) have investigated the use of dextrans
standards in a linear calibration method using two broad standards. Comparison of the calibration curve generated with this
method and a calibration curve generated by use of M_{rms} vs peak
elution volume for a series of dextran standards showed excellent
agreement. Figure 6 displays the chromatograms of the dextran
standard and T-40 dextran sample.

Column Set A

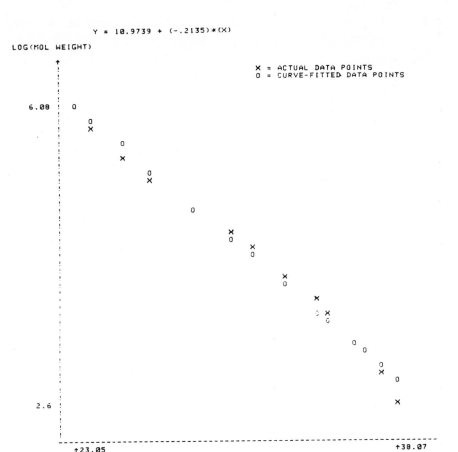

Y = 10.9739 + (-.2135)*(X)

LOG(MOL WEIGHT)

X = ACTUAL DATA POINTS
O = CURVE-FITTED DATA POINTS

ELUTION VOLUME (MLS)

VERTICALLY: EACH '!' REPRESENTS .0869 UNITS
HORIZONTALLY: EACH '-' REPRESENTS .2503 UNITS

Figure 4A. Calibration curves using a series of narrow MWD polyethylene oxide standards for MicroPak TSK PW column sets. Linear least squares fit for log (MW) vs. elution volume. Column set A: MicroPak TSK 3000PW + 4000PW + 5000PW + 6000PW.

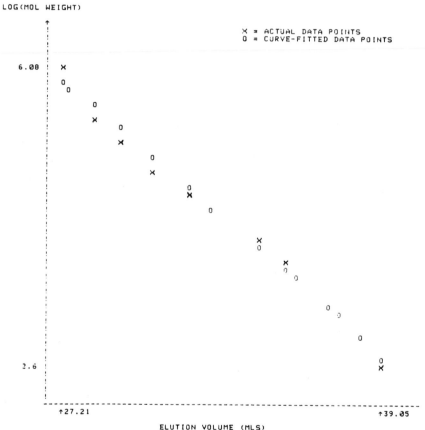

Figure 4B. Calibration curves using a series of narrow
MWD polyethylene oxide standards for MicroPak TSK PW column
sets. Linear least squares fit for log (MW) vs. elution
volume. Column set B: MicroPak TSK 3000PW + 3000PW +
6000PW + 6000PW.

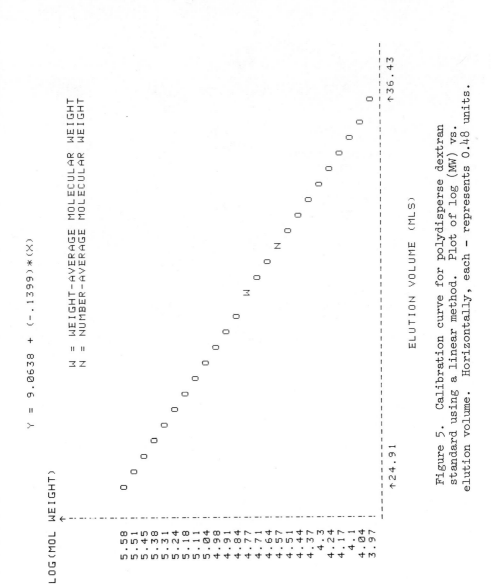

Figure 5. Calibration curve for polydisperse dextran standard using a linear method. Plot of log (MW) vs. elution volume. Horizontally, each – represents 0.48 units.

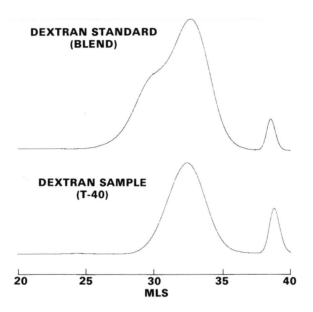

Figure 6. Chromatograms of polydisperse dextran calibration standard and dextran T-40 sample. Detector: refractive index.

Table IX. Molecular Weight Accuracy of the Linear Calibration
Method for Dextran

	$\bar{M}_w$	$\bar{M}_n$
Reported Value	39,900	26,200
Linear Method Using a Poly-		
disperse Std.	37,799	30,091
% Difference	-5.3%	+14.8%

Due to the fact that application of universal calibration is
not always practical in aqueous SEC, a linear calibration method
using a single polydisperse standard has a high degree of viabil-
ity for characterization of water-soluble polymers. Although few
water-soluble polymers with characterized MWD moments are commercially available, in many instances an in-house polydisperse
standard can be generated by measuring $\bar{M}_n$ and $\bar{M}_w$ of one lot of
polymer of the same chemical type as that under study.

Optimization of Linear Calibration Methodology. The accuracy of
linear calibration methods for utilization of polydisperse calibration standards depends upon (1) how well the column set approximates true linearity over the molecular weight calibration range
and (2) the extent to which instrumental band broadening affects
the elution volume profile of the polydisperse standard.
     Linearity of the SEC column set can be achieved by use of
commercially available mixed bed columns that have been optimized
for linearity. Alternatively, the linearity of a SEC column set
can be optimized by coupling columns of different pore sizes but
equal pore volumes (24).
     Instrumental band broadening or axial dispersion can cause
calibration errors when employing polydisperse standards. Correction of the polydisperse standard calibration data for instrumental band broadening will minimize the effect on molecular
weight analyses of polymer samples. However, as previously demonstrated in this report, when low dispersion SEC columns are
employed instrumental band broadening is minimized and the effect
on use of linear calibration methodology is negligible.

Conclusion. A linear calibration method based upon that of Balke
and Hamielec and incorporating a very efficient two variable
search algorithm was evaluated from the standpoint of calculated
molecular weight accuracy in both non-aqueous and aqueous SEC.
A comparison to calculated molecular weight accuracy with peak
position, universal calibration, and Q-factor approximation
methods using a series of narrow MWD standards was performed.
From these studies the following conclusions have been drawn:
1. The linear calibration method provides an equivalent molecular
    weight calibration curve to a peak position method of calibra-
    tion using a series of polystyrene standards.

2. Based upon studies with the NBS standard reference materials,
    the linear calibration method appears to give equivalent
    accuracy compared to a peak position method irrespective of
    sample dispersity provided that the sample elute over the
    elution volume range covered by the polydisperse standard and
    low dispersion SEC column are utilized.
3. In comparing the linear calibration method to a universal
    calibration method for PVC polymers, the universal calibration
    method appeared to be slightly more accurate suggesting that
    universal calibration methodology be applied whenever possi-
    ble. In cases where universal calibration can not be
    utilized, the linear calibration method provides a viable
    alternative.
4. The linear calibration method has utility for characteriza-
    tion of water-soluble polymers due to the constraints
    imposed in aqueous SEC towards universal calibration method-
    ology. A cursory evaluation of the linear calibration method
    for aqueous SEC indicates the method can be used with a high
    degree of accuracy to calculate molecular weight distribu-
    tion values.

Acknowledgments
The authors would like to kindly thank Denise Thomas for preparation of the manuscript.

Literature Cited

1. E.A. Collins, J. Bares, and F.W. Billmeyer, Experiments in
    Polymer Science, John Wiley and Sons Inc., New York, 1973,
    p. 312.
2. W.W. Yau, J.J. Kirland, and D.D. Bly, Modern Size-Exclusion
    Liquid Chromatography, John Wiley and Sons, New York, 1979
    Chapter 9.
3. Provder, T., Woodbrey, J.C., and Clark J.H., Separation
    Sciences, 1971, 6, p. 101.
4. Hamielec, A.E., and Omorodion, S.N.E., Size Exclusion Chroma-
    tography (GPC) , ACS SYMPOSIUM SERIES, No. 138, 1980, Chapter
    9.
5. Cantow, M.J.R., Porter R.S. and Johnson, J.F., Journal of
    Polymer Science: Part A-1, 5. 1967, pp 1391-1394.
6. Weiss, A.R., and Cohn-Ginsberg, E., J. Polymer Science, Part
    A-2, 8, 1970, p 148.
7. Swartz, T.D., Bly D.D. and Edwatds, A.S.M., J. Applied Polymer
    Science, 16, 1972, p. 3353.
8. Abdel-Alim, A.H. and Hamielec, A.E., J. Applied Polymer
    Science, 18, 1974, p. 297.
9. Frank, F.C., Ward, I.M., Williams, T., Wills, H.H., J. Polymer
    Science, Part A-2, 6, 1969, pp 1357-1369.
10. Balke, S.T., Hamielec, A.E., LeClair, B.P., and Pearce, S.L.,
    Ind. Eng. Chem. Prod. Res. Develop., 8, 1969, pp 54-57.

11. Loy, B.R., J. Polymer Science: Polymer Chem. Ed., 14, 1976, p 2321.
12. Pollock, M. MacGregor, J.F. and Hamielec, A.E., J. Liquid Chrom., 2, 1979, pp 895-917.
13. Yau, W.W., Stoklosa, H.J., and Bly D.D., J. Applied Polymer Science, 21, 1977, pp 1911-1920.
14. Malawer, E.G., and Montana, A.J., J. Polymer Science: Polymer Physics, 18, 1980, pp 2303-2305.
15. Alfredson, T.V., Perry W.J., and Tallman, L., Automated GPC Data Handling for Molecular Weight Calculations of Polymers, paper presented at 1982 Pittsburgh Conference and Exposition on Analytical Chemistry and Applied Spectroscopy, March 1982, Atlantic City, NJ.
16. Cazes, J., J. Chem. Educ., 43, p A567 - A625, 1966.
17. Grubistic, Z., Rempp, R. and Benoit, H., J. Polym. Science Part B, 5, p 753, 1967.
18. Alfredson, T.V., personal communication from Edwin R. North, Analytical Laboratories, Polymer Analysis Group, Dow Chemical USA, Midland, Michigan
19. Provder, T. and Rosen E.M., Separation Science, 5, 1970, p 437.
20. Freeman, M., and Manning, P.B., Journal of Polymer Science, Part A-2, Polymer Physics, 1964, p 2017.
21. Yau, W.W., Stocklosa, H.J. and Bly D.D., J. Applied Polymer Science, 21, 1977, p 1911.
22. Ambler, M.R., J. Polymer Science: Polymer Chem. Ed., 11, 1973, p 191.
23. Hamielec, A.E. and Omorodion, S.N.E., Size Exclusion Chromatography (GPC), ACS SYMPOSIUM SERIES, No. 138, Chapter 9, p. 193.
24. Yau, W.W., Ginnard, C.R. and Kirland, J.J., J. Chromatogr., 149, p 465, 1978.

RECEIVED October 4, 1983

Size Exclusion Chromatography of Polyethylenes
Reliability of Data

L. A. UTRACKI and M. M. DUMOULIN

National Research Council Canada, Industrial Materials Research Institute, 75 Boulevard de Mortagne, Montréal, Québec, Canada, J4B 6Y4

A reliable procedure for determination of molecular parameters: number, weight and z-averages of the molecular weight (M_i, i = n, w and z respectively) for polyethylenes, PE, by means of Size Exclusion Chromatography, SEC, has been developed. The Waters Sci. Ltd. GPC/LC Model 150C was used at 135°C with trichlorobenzene, TCB, as a solvent. The standard samples as well as commercial stabilized and not stabilized PE-resins were evaluated. The effects of: sampling, method of solution preparation, addition of antioxidant(s), thermal and shear degradation were studied. The adopted procedure allows reproducible determination of M_n and M_w, with a random error of ± 4% and M_z, with ± 9%, within 2 to 72 hrs from the initial moment of preparation of solutions.

While separation of ions according to size had already been observed by Ungerer in 1925 the first application of the principle to polymers occurred 19 years later ($\underline{1}$). Between 1960 and 1962, Vaughan and Moore ($\underline{2}$) independently developed methods for preparation of crosslinked polystyrene gel beads. The latter author is also credited with design of the analytical SEC as we know it today. Modern equipment ($\underline{3}$, $\underline{4}$) operates at higher pressure, which combined with the higher temperature required for analysis of most polyolefins, results in a drastic shortening of column life time. Tempered alkali borosilicate glasses, leached with acids to produce uniform pore size, may eventually provide a solution ($\underline{5-14}$). Unfortunately, they exhibit two disadvantages: low efficiency and solute adsorption.

Polyethylenes, PE, have been characterized by SEC since the mid-sixties and frequent problems with polystyrene gel columns have been reported ($\underline{6}$). The low density PE, LDPE, because of complexity of the molecular weight and branching distributions,

0097-6156/84/0245-0097$06.00/0
Published 1984, American Chemical Society

enjoyed   more   attention   (3,15-23)   than   the   simpler,   and
historically newer, high density, HDPE (24, 25). The results on
the ultra high molecular weight, UHMWPE, have been published
only recently (26, 27).

In this first report on SEC of PE, we want to comment on
reproducibility of the measurements. This has been discussed by
Nakajima (24) and others (28). In both cases seven PE samples
were dissolved in 1,2,4-trichlorobenzene, TCB, and tested at
130°C using 4 or 5 polystyrene-gel columns. The standard
deviations: σ_n = 6.43 and 3.4 to 5.6%, as well as σ_w =
7.43 and 3.4 to 4.4 were reported in these publications,
respectively (subscript n and w refers to number and weight
averages). The maximum spread of values Δ_n = 17.4 and
Δ_w = 25% was observed. The authors (28) reported
significant time changes in column separation properties.
Standard deviations in low temperature SEC: σ_n = 4, σ_w =
5%, were reported (29, 30).

Experimental

A Waters Sci. Ltd, GPC/LC Model 150C with Waters Model 130 Data
Module was used. The instrument was operated at 135°C with TCB
as a solvent (HPLC grade from Fisher Sci., filtered through 0.5
μm filter with silicagel). Four and five μ-Styragel (Waters
Sci. Inc.) columns with pore sizes: (500), 10^3, 10^4, 10^5,
10^6 Å (from pump to detector) were calibrated using 21
narrow MWD polystyrene samples supplied by Pressure Chemicals
and Waters Sci. (peak molecular weight M_p = 826 to 1.987 x
10^6 and polydispersity ratio M_w/M_n = 1.02 to 1.21). The
columns were calibrated (31) at 135°C using 0.06% of polymer
(three standards per solution) in TCB. The calibration was
checked once a week.

For calibration, the solutions were prepared overnight at
ambiant temperature without agitation, filtration or addition of
antioxidants (a mild agitation and filtration resulted in an
increase of retention time, RT, by 0.40 min, equivalent to a
reduction of molecular weight by 26%). The calibration curve
for the four columns Figure 1 was non-linear; addition of the
fifth, 500 Å column, Figure 2 linearized the dependence:

$$\log M_p = 11.655222 - 0.170919 \text{ RT}, \qquad 30 \leqslant RT \leqslant 55 \qquad (1)$$

with the standard error of estimate σ = 0.043816 and the
correlation coefficient r^2 = 0.99914. During the calibration,
as well as during the testing, the same conditions, listed in
 Table I , were used. Neither spinning nor filtering
operational options were used.

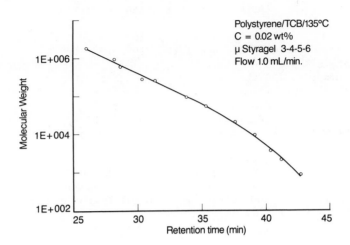

Figure 1.   Calibration curve for four μ-styragel columns
(PS in TCB at 135 °C).

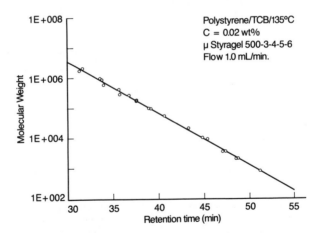

Figure 2.   Calibration curve for five μ-styragel columns
(PS in TCB at 135 °C).

TABLE I: SEC OPERATING CONDITIONS

Columns:	I- μ-Styragel; 500, 10^3, 10^4, 10^5, 10^6 Å
	II- μ-Styragel; 10^3, 10^4, 10^5, 10^6 Å
Temperature:	135°C
Inj. volume:	400 μL
Flow rate:	1 mL/min
Sen./scale:	64/25
Solvent:	1, 2, 4-trichlorobenzene (TCB)
Antioxidants:	I- Topanol (1,1,3-tri(ter-butyl hydroxy methyl phenyl)butane) and Nonox DLTDP (di laurylthiodipropionate); 0.1 wt% in solution
	II- Santonox-R (4,4'-thio-bis-(6-terbutyl m-cresol); 0.02 wt% in solvent.
Solutions:	< 0.1 wt%; (90 min. at 165°C and 30 min. at 135°C)

The PE samples were dissolved in TCB at 165°C for 1.5 hrs, then transferred to the SEC injection chamber and after 30 min injected. In the later stage of work, 0.1 wt% of two antioxidants: Topanol and Nonox DLTDP were added to the mobile phase only. The chromatograms (see Figure 3) were collected and evaluated on Waters Data Module (see Table II).

TABLE II: COMPUTATIONAL PARAMETERS

Low molecular weight limit		1500
Area rejection		700
Data collection:	first slice	21 min.
	last slice	43 min.

Commercial: high density, medium density, low density and linear low density PE's (HDPE, MDPE, LDPE and LLDPE respectively) were used. Their properties are listed in Table III .

TABLE III: THE CHARACTERISTICS OF THE COMMERCIAL PE RESINS

N°	RESIN	DENSITY ρ(Kg/L)	MELT INDEX MI(g/10min)	ZERO SHEAR VISCOSITY η_o(Pa.s)x10^{-3} 190°C	M_wx10^{-3}	M_z/M_n
1	HDPE	0.955	14	1.4	125	11
2	MDPE	0.941	0.25	74.5	325	32
3	LDPE	0.924	0.80	29.0	155	8.5
4	LLDPE	0.920	1.1	8.9	212	6.3

Most of the initial work was done using the "as received" HDPE resin. The effects of its degradation during processing and testing on the molecular weight parameters were also studied; in the text the following code for these samples will be used: V – as received, G – mixed at 210°C for 15 min on a roll mill and granulated, M–in addition to G molded at 170°C in 8 min, C and CC–in addition to M sheared for 0.5 and 1.5 hrs at 190°C and frequency ω = 0.1 (rads/s).

Results

First, a 50 ml solution was prepared of the sample V without (V-O) and with (V-A) antioxidants. The solution was poured into 12 sampling bottles and injected immediately and then every 2 hrs for 72 hrs. The variation of M_i's with time for these samples is shown in Figures 4 and 5 , respectively.

The results were fitted to the exponential relation:

$$M_i = M_{o,i} \cdot \exp\{-b_i t\} \tag{2}$$

where i = n, w and z for number, weight and z–average molecular weight, b_i is the degradation kinetics constant and t is the degradation time. The parameters of eq. (2) along with the: r^2–correlation coefficient squared, M_i–average value of the molecular weight, σ_i–standard error of the estimate and the Δ_i–maximum spread, are listed in Table IV . The polydispersity ratios: M_w/M_n or M_z/M_n did not show any time dependence. For this reason only their average values as well as σ_i and Δ_i are listed.

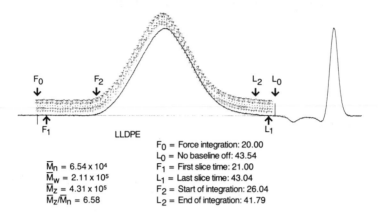

$\overline{M}_n = 6.54 \times 10^4$
$\overline{M}_w = 2.11 \times 10^5$
$\overline{M}_z = 4.31 \times 10^5$
$\overline{M}_z/\overline{M}_n = 6.58$

F_0 = Force integration: 20.00
L_0 = No baseline off: 43.54
F_1 = First slice time: 21.00
L_1 = Last slice time: 43.04
F_2 = Start of integration: 26.04
L_2 = End of integration: 41.79

Figure 3. Example of chromatogram with the indicated location of the computational parameters.

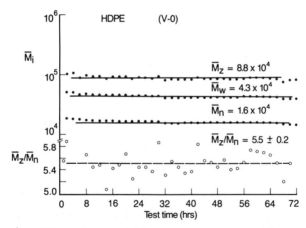

Figure 4. Molecular weight averages vs. residence time at 135 °C; HDPE without antioxidant.

TABLE IV: STATISTICS OF SEC MEASUREMENTS OF HDPE
WITHOUT (I) AND WITH (II) ANTIOXIDANT

PARAMETER	$M_{o,i}$	b_i	r	M_i	$\pm\sigma_i(\%)$	$\Delta_i(\%)$
I without antioxidants (V-O)						
1. M_n	17,510	39.32	0.7484	16,055	5.99	28.0
2. M_w	46,845	101.27	0.6304	43,098	6.15	31.1
3. M_z	95,321	181.44	0.4429	88,312	6.70	39.4
4. M_w/M_n	------	------	------	2.68	1.49	5.4
5. M_z/M_n	------	------	------	5.50	3.45	17.8
II with antioxidants (V-A)						
1. M_n	36,777	34.06	0.2480	35,483	3.89	16.3
2. M_w	122,934	-56.37	0.0685	125,076	3.48	18.7
3. M_z	391,302	-806.41	0.1659	421,945	9.48	47.5
4. M_w/M_n	-------	-------	------	3.53	4.25	14.6
5. M_z/M_n	-------	-------	------	11.91	9.99	41.6

Next, seven randomly selected pellets of the resin V were dissolved in seven bottles and injected after 6 hrs at 135°C. The solutions contained the two antioxidants. The statistics are shown in Table V.

TABLE V: EFFECT OF PE SAMPLING ON SEC DATA

PARAMETER	M_i	$\pm\sigma(\%)$	$\Delta\%$
M_n	33,135	6.88	21
M_w	130,527	7.46	25
M_z	452,356	15.79	56
M_w/M_n	3.57	5.55	--
M_z/M_n	12.51	14.10	--

Similarly, the processed samples (V through to CC) were each dissolved and injected after 2 to 17 hrs at 135°C. The average values of molecular parameters are given in Table VI . These solutions contained the two antioxidants. The standard deviations varied from the minimum values of 1.72, 1.51 and 2.25 for M_n, M_w and M_z of sample C, respectively to the maximum values of 5.99, 6.15 and 6.70 recorded for sample V.

TABLE VI: EFFECT OF PROCESSING OF PE

HDPE CODE	$t_{190}(min)$	M_n	M_w	M_z	M_z/M_n
V	0	35,962	122,252	397,110	10.99
G	25.64	40,216	130,913	393,728	9.81
M	30.10	37,718	124,380	383,230	10.18
C	60.10	38,009	123,922	357,619	9.41
CC	120.10	36,458	119,602	337,851	9.28

To check on the general applicability of the method the remaining MDPE, LDPE and LLDPE were tested using the same experimental procedure. The results are shown in "Figures 6 to 8", respectively.

Discussion

As seen in Table IV the M_n, M_w and M_z for the stabilized solutions of the HDPE sample are larger than those for the unstabilized ones by a factor of 2.2, 2.9 and 4.8 respectively. Assuming that this variation is due to thermal degradation during the dissolution and testing one can calculate the activation energy E_a = 62.5 (kcal/mole). This value compares well with E_a = 52.6 to 66.1 determined (32) at T = 375 to 436 (°C) for HDPE of molecular weight of 16 to 23 thousand, respectively. The initial results, and those collected after prolonged storage in the injection chamber, were not consistent with those collected within the "stable period": 4≤t≤68 hrs. This fact was further demonstrated in another series of measurements where the samples were injected for 230 hrs. The initial values of M_i widely scattered, whereas those for t>68 hrs systematically increased with time (this increase is responsible for the negative values of b_w and b_z in Table IV). Apparently, dissolving HDPE sample at 165°C for the period of 1.5 hrs is not sufficient. Only after an additional 2.5 hrs in the injection chamber at 135°C is the

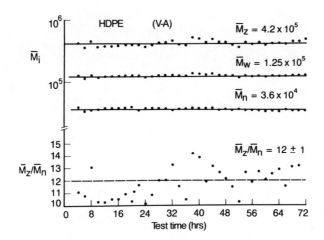

Figure 5.   Molecular weight averages vs. residence time at 135 °C; HDPE with antioxidants.

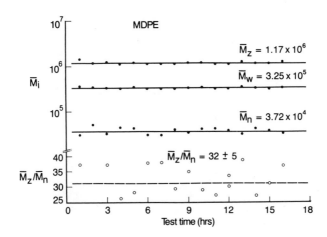

Figure 6.   Molecular weight averages vs. residence time at 135 °C; MDPE with antioxidants.

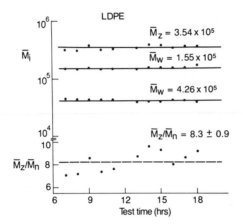

Figure 7. Molecular weight averages vs. residence time at 135 $^\circ$C; LDPE with antioxidants.

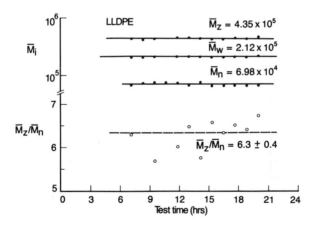

Figure 8. Molecular weight averages vs. residence time at 135 $^\circ$C; LLDPE with antioxidants.

dissolution process completed and the stable, reproducible values of M_i's are obtained. On the other hand, prolonged heating of the sample in the presence of antioxidants leads to gradual increase of M_i's for times $t_c \geqslant t \geqslant 32$ hrs. The value of t_c was observed to vary from one resin to another. It is worth noting that r^2 for M_i in V-A series is very low, indicating a random variation. The standard deviations of these data $(0 \leqslant t \leqslant 72)$ are $\sigma_n = \sigma_w = 4\%$ and $\sigma_z = 9\%$, which compare quite well with the previously quoted literature results.

It has been reported (33) that MWD of HDPE can be described by the log-normal distribution function (34):

$$p(x) = [\sigma(2\pi)^{1/2}]^{-1} \exp \{-t^2/2\} \qquad (3)$$

$$t = (x-\bar{x})/\sigma$$

where $x = \log M$, $\bar{x}$ is the mean value of x and σ is the standard deviation. Defining the normal equivalent deviate as a proportion of $p(x)$ which exceeds the integral:

$$p(s) = (2\pi)^{-1/2} \int_{-\infty}^{s} \exp \{-t^2/2\} \, dt \qquad (4)$$

one can conveniently plot the probability function as p vs. probit, where probit is taken as $(s+5)$.

The V-A data follow Equation 3 quite well, with $x = \log M_n$, $M_n = 41,527 \pm 878$ and $\sigma = 1.585 \pm 0.005$. On the other hand, the V-O data cannot be represented by this function. One can postulate that PE in TCB undergoes a random scission similar to that observed for polymer melts at much higher temperatures. In Figure 9 the integral distribution curves of samples HDPE (V-0) and (V-A), both taken after 10 hrs of dissolution, are shown in the form of log-normal distribution plot: M vs probits. Two facts are apparent: (1) the molecular weights of sample V-A (broken line) are systematically higher than those of V-0 (points); (2) when the V-A distribution curve is displaced vertically to coincide with that of V-0 in the region of low molecular weight, it is quite apparent that the degradation preferentially affected the molecules with $M > 10^5$, while below this value $M_{V-A} \approx KM_{V-0}$, with K being a constant, $K \approx 2.2$. For $M > 10^5$ K increases with molecular weight approximately as: $K = 2.2 + 2.1 \log M$.

The above analysis should not be construed as authors' opinion that molecular weight distribution, MWD, of PE's should follow log-normal probability. The method of analysis is general and does not require that Equation 3 be obeyed; if it does, log M vs. probit is a straight line, which simply makes the

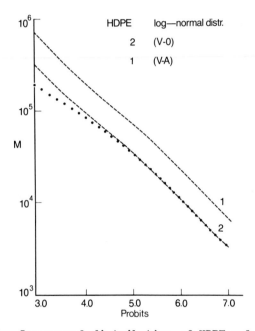

Figure 9. Log-normal distribution of HDPE molecular
weight after 1.5 h at 165 °C and 10 h at 135 °C; 1 (upper
line) - HDPE with antioxidant; circles - HDPE without antioxidant; 2 (lower line) upper broken line has been deplaced
vertically by a factor of 2.2 to coincide with the low
molecular weight data of unstabilized sample.

work a little easier. We found that the plot is useful in interpreting the data even in the case of multimodal distributions.

The commercial resins are seldom a result of a single polymerization; in order to meet the specifications they are blended. The results of Table V show that the variability of M_i's in this series is larger than that in Table IV ; in the first case the results refer to average values for seven different pellets of HDPE-V dissolved separately, in the second to a variability of data of the same solution. The statistical analysis of the first set of data indicates that there is about 9% pellet-to-pellet variability in M_w.

In Table VI the results of SEC-testing of the processed samples are shown. The t_{190} represents the equivalent degradation time at 190°C calculated from the actual times and temperatures reported in the table; in the calculation, a simple Arrhenian function was assumed, with the activation energy ΔE_a = 11.9 (kcal/mole) obtained during the previous work (35). In Figure 10 the M_i's dependence on t_{190} is shown. The results are most encouraging. It can be seen that even prolonged heating of the resin, under processing conditions, does not lead to a significant alteration of its M_i's. The onset of degradation becomes apparent only for sample CC; here M_z is 17.5% lower than that of sample sample V. Since standard deviation of the measurements is ± 9.5% the drop in M_z reflects the true degradation. This is more clearly visible on Figure 10 , where the polydispersity parameter, M_z/M_n, decreases systematically from a value of about 11 to 9.3. The initially more rapid decrease of this parameter is most likely due to the easy access of oxygen during this stage of the process - a factor neglected in calculating t_{190}.

Finally, a few words on the general reliability of the developed method of the measurements. The method, as shown in Figures 4 and 6 to 8 works quite well for all PE's of a normal, commercial range of M_i's. We observed a need for longer dissolution time of HDPE than that of LDPE or LLDPE of equivalent molecular weight. The adopted dissolution time is 1.5 hrs at 165°C and 2.5 hrs at 135°C. With the weekly recalibration procedure the long term repeatability of data during the two years period was found to be random, and within the range of the reported standard deviations. Some initial work on SEC of the UHMWPE has been conducted; it was found that the above conditions were grossly inadequate.

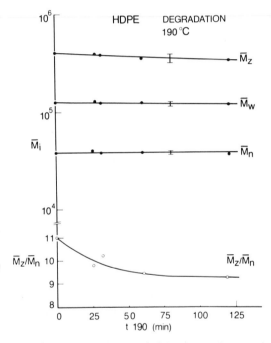

Figure 10.   Molecular weight parameters of HDPE vs. the
processing time at 190 °C; see text.

Acknowledgments

The authors would like to thank Mr. J. Dufour for his careful work in collecting the SEC data.

Literature Cited

1. J. Claesson and S. Claesson, Arkiv. Kemi, 19A, 1 (1944).
2. J.C. Moore, J. Polymer Sci., A2 835 (1964).
3. E.E. Drott and R.A. Mendelson, J. Polymer Sci., Part A-2, 8, 1361, 1373 (1970).
4. Polymer Laboratories "PLgel GPC columns" technical buletin 5/82.
5. W. Haller, Nature, 206, 693 (1965).
6. J.H. Ross and M.E. Casto, J. Polymer Sci., Part C, 21, 143 (1968).
7. A. Titterton, Ind. Polym., May 1-2, 1973 pg. 83-88.
8. A.R. Cooper, J. Polymer Sci., Polymer Phys. Ed., 12, 1969 (1974).
9. L. Westerman, Chromatog. Sci., 13, 257 (1980).
10. M. Kubin, J. Appl. Polymer Sci., 27, 2933, 2955 (1982).
11. B.W. Hatt, Develop. Chromatogr., 1, 157 (1978).
12. J.V. Dawkins and G. Yeadon, Develop. Polym. Charact., 1, 71 (1978).
13. J.V. Dawkins, Pure Appl. Chem., 51, 1473 (1979).
14. R.A. Ellis, Pigm. Res. Techn., Sept. 1979, pg. 10-21.
15. Z. Grubisic, P. Rempp and H. Benoit, J. Polymer Sci., B5, 753 (1967).
16. B.H. Zimm and W.H. Stockmayer, J. Chem. Phys., 17, 1301 (1949).
17. J.A. Cote and M. Shida, J. Polymer Sci., Part A-2, 9, 421 (1971)
18. J.A. Miltz and A. Ram, Polymer, 12, 685 (1971); A. Ram and J. Miltz, J. Polymer Sci., 15, 2639 (1971).
19. G.R. Williams and A. Cervenka, Eur. Polymer J., 8, 1009 (1972).
20. S. Nakano and Y. Goto, J. Appl. Polymer Sci., 19 2655 (1975); ibid., 20, 3313 (1976).
21. L. Wild, R. Ranganath and A. Barlow, J. Appl. Polymer Sci., 21, 3319, 3331 (1977).
22. L. Lecacheux, J. Lesec and C. Quivoron, ACS Polymer Prepr., 23(2) 126 (1982).
23. A. Hamielec, Pure Appl. Chem., 54, 293 (1982).
24. N. Nakajima, J. Appl. Polymer Sci., 15, 3089 (1971); idid. 16, 2417 (1972).
25. J.V. Dawkins and J.W. Maddok, Eur. Polymer J., 7, 1537 (1971).
26. A. Barlow and T. Ryle, Plastics Eng., August 1977, pg. 41-43.

27. Polymer Laboratories, technical information 5/82.
28. G. Samay and L. Fuzes, J. Polymer Sci., Polymer Symp., 68, 185 (1980).
29. J.H. Duerksen and A. Hamielec, ACS Symp. on Analytical GPC, Chicago, Sept. 1967.
30. J.P. Busnel, Polymer, 23, 137 (1982).
31. A.E. Hamielec and A.C. Ouano, J. Liq.Chromatography, 1, 111 (1978).
32. H.H.G. Jellinek, J. Polymer Sci., 4, 13 (1949).
33. H. Wesslun, Makromol. Chem., 20, 111 (1956).
34. W.D. Lansing and E.O. Kraemer, J. Am. Chem. Soc., 57, 1369 (1935).
35. L.A. Utracki and J. Lara, Proceedings of the Int'l Workshop on Extensional Flows, Mülhouse – La Bresse, France, 24-28 January 1983.

RECEIVED September 12, 1983

Gel Permeation Chromatography
Correction Procedure for Imperfect Resolution

B. A. ADESANYA, H. C. YEN, and D. C. TIMM

University of Nebraska, Lincoln, NE 68588-0126

N. C. PLASS

Brunswick Corporation, Lincoln, NE 68504

In part I, Timm and Rachow (1) describe an algorithm for
interpretation of chromatograms for imperfect resolution. The
instrument was one of low plate counts, and yet population density distributions consistent with theoretical, kinetic models
were achieved (2,3). Research, using high plate count columns,
shows that convergent distributions are achieved and that results
are not a function of instrument resolution. Linear polystyrene
resins had a polydispersity in the interval $1.5 \leq \overline{M}_w/\overline{M}_n < 2.0$.
Data analysis includes mass fractions of unreacted monomers and
species of similar molecular weight.

A second algorithm is described for analysis of resins of
narrow, molecular distributions $\overline{M}_w/\overline{M}_n \simeq 1.0$. Experimental testing incorporates polystyrene initiated with n-butyl lithium
and a linear, step-growth epoxy comprised of nadic methyl anhydride and phenyl glycidyl ether. Kinetic distributions are described by a Poisson molar distribution. The accuracy of experimental population density distributions for macromolecular species is observed to be limited by the precision of the average
molecular weights determination by light scattering and by vapor
pressure osmometry. The algorithm may be constrained to fit a
Poisson molar distribution. Experimental error is more pronounced for higher molecular weight resins, which require greater
precision in assignments of molecular weights.

Calibration

The algorithm calibration sequence is pictorially shown in
Figure 1. Chromatograms for monomers plus polymers of narrow,
molecular distribution are experimentally observed, normalized to
a unit mass and labeled S_{ij}. The index i identifies equally-
spaced, elution volume increments from time of sample injection;
the index j defines the standard number. For each polymeric
standard, the molecular weight of species eluting in the volume
interval VE_i is assumed to be a semilogarithmic function of
elution volume

NOTE: This is Part II in a series.

0097–6156/84/0245–0113$06.00/0

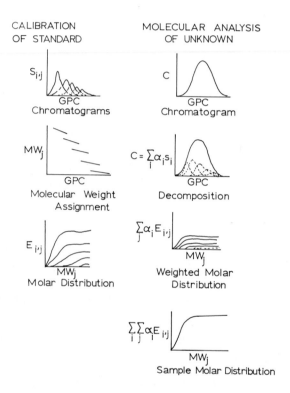

Figure 1.   Schematic for GPC calibration and sample analysis.

$$\ln MW_{ij} = A_j + B_j(VE_i) \tag{1}$$

Analysis of resins described by Poisson distributions shows the validity of this constraint for standards. A cumulative, molar distribution of macromolecules is determined from the normalized chromatogram S_{ij} and Relationship 1:

$$E_{nj} = \sum_{i=1}^{n} S_{ij}/MW_{ij} \tag{2}$$

Degree of polymerization is n. Molar distributions are interpolated to specific degrees of polymerization.

In evaluation of the parameters A_j and B_j, Timm and Rachow (1) utilized number and weight average molecular weights coupled with the observed, normalized chromatogram for that standard. Specifically,

$$PD_j = \sum_i (S_{ij} \exp(B_j VE_i)) \sum_i (S_{ij} \exp(-B_j VE_i)) \tag{3}$$

A Newton/Raphson iteration yields the value for B_j. An average molecular weight yields the value of A_j. In reference to Figure 1, the calibration sequence is now completed. Block data storage incorporates S_{ij} and E_{nj}.

Population Density Distributions

Figures 2 and 3 present typical results obtained from a low plate count column and a high plate count column. The graphs present the calculated molar concentrations of macromolecular species as a function of their degree of polymerization. The straight lines are the theoretical, kinetic distributions. Inasmuch as convergent solutions are obtained, the algorithm is effective for correction for imperfect resolution.

The polystyrene data were collected from a steady state, continuous, well-mixed reactor. The initiator was n-butyllithium for data of Figure 2 and was azobisisobutylnitrile for data of Figure 3. Toluene was used as a solvent. The former polymerization yields an exponential population density distribution (2), $\overline{M}_w/\overline{M}_n = 1.5$; the latter yields a molar distribution defined as the product of degree of polymerization and an exponential (3), $\overline{M}_w/\overline{M}_n = 2.0$. Standards utilized in calibration of both instruments were polystyrene supplied by Pressure Chemical Company.

Poisson Distribution

For a polymerization comprised of propagation kinetics only, a Poisson molar distribution exists for a batch polymerization initially seeded with a polymeric species $A_1(0)$. Rate of propagation is defined by

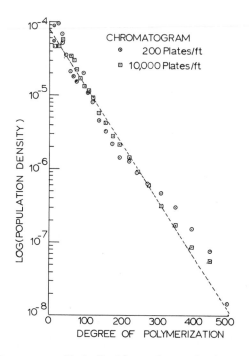

Figure 2.   Frequency distributions for polystyrene initiated
with n-BuLi.

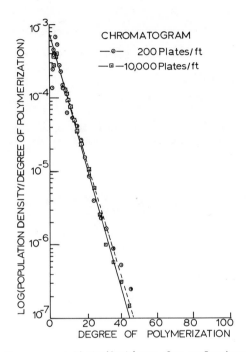

Figure 3. Frequency distributions for polystyrene initiated with AIBN.

$$A_n + M \rightarrow A_{n+1} \qquad K_p$$

The molar concentration of a polymeric molecule of degree of polymerization n is expressed as

$$\frac{dA_n(t)}{K_p M(t)dt} = \frac{dA_n(t)}{d\tau} = A_{n-1}(t) - A_n(t)$$

Seeding yields a null initial condition except for

$$A_1(0) \neq 0$$

Integration yields a Poisson, molar frequency distribution

$$A_n(\tau) = A_1(\tau)\ \tau^{n-1}\ \exp(\tau)/(n-1)! \tag{4}$$

The weight distribution is

$$W_n(\tau) = M_o\ n\ A_n(\tau) \tag{5}$$

The molecular weight of the repeat unit is M_o. The number average and weight average degrees of polymerization are

$$\overline{DP}_n = \frac{\overline{M}_n}{\overline{M}_o} = \frac{\sum\limits_{k=0}^{\infty} k^1 A_k(\tau)}{\sum\limits_{k=0}^{\infty} k^o A_k(\tau)} = 1 + \tau \tag{6}$$

$$\overline{DP}_w = \frac{\overline{M}_w}{\overline{M}_o} = \frac{\sum\limits_{k=0}^{\infty} k^2 A_k(\tau)}{\sum\limits_{k=0}^{\infty} k\ A_k(\tau)} = 1 + \tau + \frac{\tau}{1+\tau} \tag{7}$$

Hence, polydispersity is

$$PD = \frac{\overline{DP}_w}{\overline{DP}_n} = 1 + \frac{\tau}{(1+\tau)^2} \geq 1.0 \tag{8}$$

To experimentally control the average molecular weight for such polymerizations, the initial concentration of seeds $A_1(0)$ relative to monomer $M(0)$ is manipulated. The lower this ratio, the higher will be the ultimate average molecular weight, which, in turn, increases the value of τ, the integral conversion of monomer. Chain-growth polymerizations of styrene, initiated with n-butyllithium approximate such a kinetic mechanism (4). The step-growth polymerization of the following epoxy resins will also yield a Poisson, molar, frequency distribution.

Catalyst	Basic tertiary amine
Seed	Benzoic acid
Monomers	Phenyl glycidyl ether
	Nadic methyl anhydride

The reactive hydrogen site supplied by the organic acid controls the number of polymer molecules; the basic catalyst effectively results in the alternate addition of oxirane/anhydride monomers, forming ester linkages at the reactive hydrogen site (5).

The algorithm utilizes block data sets for S_{ij} of actually observed chromatograms and regenerates an unknown sample's chromatogram C_i by

$$C_i = \sum_j \alpha_j S_{ij} + \epsilon_i \qquad (9)$$

The weight fraction for each standard is α_j and the error in the fit is ϵ_j. Sufficient standards must be run so as to reconstruct the chromatogram C_i. Experience suggests that about fifteen are normally adequate for broadly distributed resins, i.e. $\overline{M}_w/\overline{M}_n \geq 1.5$.

When a polymer of narrow distribution is subjected to analysis, its chromatogram may fall between those for two adjacent standards. A least squares fit then yields a weighted bimodal distribution. If the chromatogram for the unknown coincides with that for a standard, the calculated distribution will be that of the standard (see Figure 1). The former results in broadening of the numerical results; the latter is desirable, but unlikely. Experience has, therefore, resulted in the development of a subroutine for such analyses. Adesanya (6) selected the observed chromatogram to be S_{ij} and initially explored average molecular weights to evaluate the parameter τ of the Poisson distribution, Equations 6 and 7. Normally reported values yield a degree of uncertainty in its numerical value assignment which becomes more significant as molecular weight increases. Thus, the constraint of Equation 3 was modified, particularly for higher molecular weight standards.

Poisson distributions exhibit a maximum near $\tau = j-1$ (7). The chromatogram's maximum, coupled with an overall calibration (the logarithm of the average molecular weight vs eluent volume at the chromatogram peak), was utilized to assign the value for τ for the unknown. The weight average molecular weight was utilized. Relationships 4 and 5 were then utilized to evaluate the theoretical, kinetic, weight distribution and correlated as a function of degree of polymerization n.

The area of the chromatogram for the unknown sample can also be utilized to generate a weight fraction distribution, but as a function of eluent volume, i (see Figure 4). At a constant mass fraction, the two distributions are equal and can be utilized to generate a calibration curve to check the validity of the semilogarithmic calibration constraint, Equation 1. Figure 5 pre-

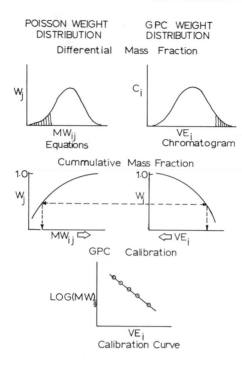

Figure 4.   Schematic for GPC calibration subject to Poisson constraints.

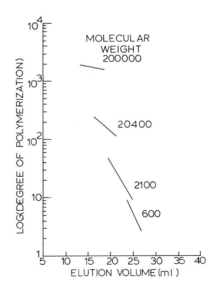

Figure 5.   Individual standard's calibration curve subject to Poisson constraints.

sents results for four polystyrenes. A semilogarithmic relation
is generated. Therefore, Equation 1 is a valid relationship for
evaluation of the cummulative, molar distribution during calibration.

As the molecular weight of the material increases, the polydispersity of that material must approach 1. Specifically, for
the 200,000 molecular weight sample, light scattering and vapor
pressure osmometry yields a polydispersity value near 1.06. If
the material is distributed according to a Poisson distribution,
the polydispersity will be 1.0005. Normal errors reported in
measurements of average molecular weights preclude this accuracy.
However, if one believes an average molecular weight is correct
and forces the second average to be consistent (Equation 6 or 7),
the calibration procedure described by Timm and Rachow, Equation
3, will yield a Poisson molecular frequency distribution. If both
experimentally observed molecular weights are utilized, the calculated distribution is normally a broader distribution than is the
Poisson distribution.

Monomer Analysis

Yen (8) and Tien (9) utilized vapor pressure osmometry and
light scattering for molecular weight analysis of linear, epoxy
resins. These have subsequently been extensively utilized by our
research group as calibration standards in the analysis of thermoset, epoxy resins (10). To obtain better estimates, chromatography was utilized to correct observations for monomer contamination. The algorithm was modified such that the originally observed chromatogram C_i was expressed in terms of standards S_{ij} by:

$$C_i = \sum_{j=1}^{m1} \alpha_j S_{ij} + \sum_{j=m1}^{m2} G_{ij} + \sum_{j=m2}^{N} \alpha_j S_{ij} + \varepsilon_i$$

The first summation incorporates a block data set for observed
monomeric standards; the second is a null buffer; the third are
polymeric standards. Testing through the addition of monomers to
polymer standards verified that the decomposition concept is
again valid. Accuracy within the chromatogram is the error-deter-
mining step. Table I presents analysis of material after blending
known quantities.

The polymerizations were designed such that formulation,
coupled to stoichiometry, would control the ultimate molecular
weight. Table II presents theoretical, kinetic, average molecular
weights, as well as those initially determined. Low molecular
weight measurements were acceptable, but high molecular weights
were in serious error, due to monomer contamination. The observed number average molecular weight can be expressed by

$$\overline{M}_n \text{ observed} = \frac{\Sigma \text{ gram(monomer)} + \Sigma \text{ gram(polymer)}}{\Sigma \text{ moles(monomer)} + \Sigma \text{ moles(polymer)}} \tag{10}$$

Table I: Decomposition of Epoxy Resin Plus Monomer,
Mass Fractions

Matl	$\overline{MW}_n$	Monomer only		Resin plus Monomer	NMA	Resin plus Monomer	PGE
		true	observed	true	observed	true	observed
BDMA	135	0.08	0.053	.023	.028	.024	.025
PGE	150	0.30	0.317	.029	.020	.121	.119
NMA	178	0.56	0.554	.174	.166	.074	.085
ROH	130	0.06	0.076	.016	.022	.016	.010
Polymer	500	0.00	0.000	.758	.764	.765	.760
Total		1.00	1.000	1.000	1.000	1.000	1.000

Table II: Epoxy Average Molecular Weights Due to
Monomeric Contamination

Kinetic* $\overline{MW}_n$	$\overline{MW}_n$		$\overline{MW}_w$		$\overline{PD}$
	observed**	corrected	observed***	corrected	corrected
330	470	490	495	500	1.02
490	650	710	1090	1100	1.54
1400	1000	1730	1630	1850	1.07
6560	2050	7600	7770	8200	1.08
328000	2250	----	21600	25700	----

*Grams monomer/mole initiator; **Vapor pressure osmometry;
***Light scattering

Chromatography analysis yields the mass of monomeric species and polymer in the cured resin, from which the moles of monomer are readily calculated. This relationship, coupled with the observed molecular weight, can be utilized to determine the moles of polymer. Hence, the corrected number average molecular weight is

$$\bar{M}_n \text{ correct} = \frac{\Sigma \text{ gram(polymer)}}{\Sigma \text{ moles(polymer)}} \tag{11}$$

Column 2 of Table II presents these results. Columns 4&5 represents corrected weight average molecular weights. For the highest molecular weight standard, the moles of monomer compared to polymer in Equation 9 are such that the correction procedure failed. However, this sample could be cleaned of monomer by fractionation techniques without serious danger of removing significant quantities of oligomers. Alternately, an analysis based solely on chromatography will yield a definitive description of the macromolecular content.

Discussion

The algorithm accurately determines the monomeric and polymeric fractions plus population density distributions of macromolecules within an unknown sample, from which mass distributions and moments may be calculated. The modified algorithm is shown to accurately evaluate resins for which $1.0 \leq \bar{M}_w/\bar{M}_n \leq 2.0$, subject to assignments of average molecular weights. Resins with polydispersity greater than 2.0 may also be evaluated, yielding expected, theoretical distributions (10). The data of Table II show that at high concentrations of initiator ROH, some species do not initiate the polymerization process. Similarly at low initiator concentrations, the achieved molecular weight is less, perhaps due to other sources of polymerization sites in the resin, one of which could be moisture.

The chromatograph is interfaced with a Digital LSI-11 microprocessor. Calibration, though requiring observations of chromatograms for monomeric and polymeric standards, can be efficiently achieved. Research over several years shows that styragel/microstyragel columns are very stable under continuous utilization.

A substantial asset of the algorithm is its flexibility, allowing for simultaneous, analytical analysis of monomeric and polymeric species, both in terms of average molecular weights and population density distributions of constitutive molecules. The technique is being extended to the analysis of extracts of quality cured thermoset resins. Simultaneous analysis by chromatography and by dynamic mechanical spectroscopy shows that oligomeric fractions' average molecular weights by chromatography closely correlate with crosslink average molecular weight determined by spectroscopy (11). Resins are of constant chemical composition, but are subjected to molecular variance through controlled cure

cycles. The technique has also successfully determined theoretical, kinetic distributions of constitutive molecules for a variety of thermoplastic resins, using a broad spectrum of polymerization mechanisms.

Acknowledgments

Financial and technical support from the Engineering Research Center and from Brunswick Corporation are appreciated.

Literature Cited

1. Timm, D.C.; Rachow, J.W. J. Polym. Sci. 1975, 13, 1401.
2. Timm, D.C.; Kubicek, L.F. Chem Engr. Sci. 1974, 29, 2145.
3. Scamehorn, J.F.; Timm, D.C. J. Polym. Sci. 1975, 13, 1241.
4. Hsieh, H.L. J. Polym. Sci. 1965, A3 153, 163, 173, 181, 191.
5. Chen, C.P.; Timm, D.C.; Ho, V. Proc. First Int'l Conf. on
   Reactive Processing of Polymers 1980, II, 4.
6. Adesanya, B.A., M.S. Thesis, University of Nebraska, Lincoln,
   NE, 1978.
7. Weast, R.C.; Selby, S.M., Ed. CRC Handbook of Tables for
   Mathematics CRD Press: Cleveland, 1975, 952.
8. Yen, H.C., M.S. Thesis, University Nebraska, Lincoln, NE,
9. Tien, C.S., M.S. Thesis, University Nebraska, Lincoln, NE,
   1980.
10. Ayorinde, A.J.; Lee, C.H.; Timm, D.C.; Humphrey, W.D., sub-
    mitted to ACS Symposium Series (1983).
11. Timm, D.C.; Ayorinde, A.J.; Huber, F.K.; Lee, C.H.; submit-
    ted to Int'l Rubber Conf. '84, 4-8 Sept., '84, Moscow.

RECEIVED September 12, 1983

Size Exclusion Chromatography Molecular Weight Separation and Column Dispersion

Simultaneous Calibration with Characterized Polymer Standards

RONG-SHI CHENG and SHU-QIN BO

Changchun Institute of Applied Chemistry, Academia Sinica, Changchun, Jilin, People's Republic of China

With the aid of the theoretical relationship between the calibration relation of a SEC column for the monodisperse polymer species under ideal working conditions and the effective relations between the molecular weight and the elution volume for characterized polymer samples, a computational procedure for simultaneous calibration of molecular weight separation and column dispersion is proposed. From the experimental chromatograms of narrow MWD polystyrene standards and broad MWD 1,2-polybutadiene fractions the spreading factors of a SEC column was deduced by the proposed method. The variation of the spreading factor with the elution volume is independent upon the polymer sample used.

A number of computer searching methods for estimating the molecular weight calibration curve of SEC with characterized polydisperse polymer standards had been proposed (1-9). Recently it has been shown that the calibration curve for a SEC column and the calculated effective relation or experimental relation between the molecular weight and the elution volume for a sample are quite different (10,11) and it is possible to estimate the molecular weight calibration curve and the spreading factor simultaneously by coupling SEC with LALLS (12). In this paper, a simple digital searching method is proposed for calibrating the molecular weight separation and column dispersion of a SEC column simultaneously with characterized polymer samples.

Theory

The molecular weight calibration function $M(V_R)$ of a SEC column may be defined as the relationship between the molecular weights of the monodisperse polymer species and their retention volume V_R under ideal working condition, i.e. in the absence of instrumental spreading effect. It is unique for a given column and the true weight and number average molecular weight of any polydis-

0097–6156/84/0245–0125$06.00/0
© 1984 American Chemical Society

perse polymer sample may be calculated by definition as

$$\langle M \rangle_w = \int W(V_R) M(V_R) \, dV_R \tag{1}$$

$$\langle M \rangle_n = 1 \, / \int (W(V_R)/M(V_R)) \, dV_R \tag{2}$$

where $W(V_R)$ is the true chromatogram of the sample. In a real SEC column, the experimental chromatogram $F(V)$ of a sample is broadened by the instrumental spreading effect and the molecular weights calculated by Equation 1 and 2 using $F(V)$ instead of $W(V_R)$ differ from the true values. We may define an effective relation between the molecular weight and elution volume $M^*(V)$ so that the true average molecular weights also satisfy the following relations:

$$\langle M \rangle_w = \int F(V) M^*(V) \, dV \tag{3}$$

$$\langle M \rangle_n = 1 \, / \int (F(V)/M^*(V)) \, dV \tag{4}$$

The effective relation $M^*(V)$ is not unique to a given SEC column but varies with samples and also differs from the calibration relation $M(V_R)$.

For a linear SEC column, the monodisperse calibration relation and the effective relations may be represented by

$$M(V_R) : \qquad \ln M = A_m - B_m V_R \tag{5}$$

$$M^*(V) : \qquad \ln M = A_m^* - B_m^* V \tag{6}$$

respectively. By using the results of the moment analysis of Tung's integral equation of instrumental spreading (13), the effective relation of a polydisperse sample may be written as (10,11)

$$M^*(V) : \qquad \ln M = (A_m - (1 - \xi)B_m \overline{V}) - \xi B_m V \tag{7}$$

where $\overline{V}$ is the mean elution volume of $F(V)$ and ξ is a parameter defined as

$$\xi^2 = (\sigma_T^2 - \langle \sigma_0^2 \rangle) \, / \, \sigma_T^2 \tag{8}$$

in which σ_T^2 is the variance of $F(V)$ and $\langle \sigma_0^2 \rangle$ is the average spreading factor of the polydisperse sample exerted on the column as expressed by

$$\langle \sigma_0^2 \rangle = \int W(V_R) \, \sigma_0^2(V_R) \, dV_R$$

The spreading factor σ_0^2 is the variance of the chromatograms of the monodisperse polymer species, i.e. of the instrumental spreading function $G(V,V_R)$. If σ_0^2 varies linearly with the retention volume of the monodisperse polymer, then $\langle \sigma_0^2 \rangle$ is numerically equal to the interpolated value $\sigma_0^2(\overline{V})$ of the function $\sigma_0^2(V_R)$ for the polydisperse sample at its mean elution volume.

It can be seen from Equation 5 and 7 that the effective relation $M^*(V)$ of a sample crosses with the unique calibration relation of the column at the mean elution volume $\bar{V}$ of that sample . After the effective relations of several samples have been deduced, the molecular weight of each sample at its cross-point may be calculated by Equation 6 and the line connecting all the crosspoints is just the calibration relation $M(V_R)$ of the column. The calibration relation may be linear or otherwise nonlinear. For the latter case the coordinates of the crosspoints may be fitted by a polynomial and then Equation 5 should be regarded as the tangent line of the polynormial which varies with the mean elution volume of the sample.

Comparing the coefficients of Equation 6 with that of Equation 7, we get

$$A_m^* = A_m - (1 - \xi) B_m \bar{V} \tag{9}$$

$$B_m^* = \xi B_m \tag{10}$$

The parameter ξ of a sample could be deduced from the slope or intercept of the effective relation and the calibration relation or its tangent and thereafter the spreading factor $\langle \sigma_\xi^2 \rangle$ could be determined from ξ by Equation 8 .

With the procedure outlined above, simple programs of programmable calculator (TI 59) and microprocessor (Z80) for finding $M(V_R)$ and $\sigma_0^2(V_R)$ were written. The mean elution volume and total variance of the experimental chromatograms of well characterized polymer samples are first calculated according to

$$\bar{V} = \Sigma H_i V_i / \Sigma H_i \tag{11}$$

$$\sigma_T^2 = \Sigma H_i V_i^2 / \Sigma H_i - \bar{V}^2 \tag{12}$$

where H_i is the height of the chromatogram at elution volume V_i .

Next the coefficients of the effective relation of each sample with known weight and number average molecular weight are evaluated by iteration. Combining Equation 3,4 and 6, the average molecular weights and inhomogeneity index may be expressed as

$$\langle M \rangle_w = \mathrm{Exp}(A_m^*) \Sigma H_i \mathrm{Exp}(-B_m^* V_i) / \Sigma H_i \tag{13}$$

$$\langle M \rangle_n = \mathrm{Exp}(A_m^*) \Sigma H_i / \Sigma H_i \mathrm{Exp}(B_m^* V_i) \tag{14}$$

$$D = \langle M \rangle_w / \langle M \rangle_n$$

$$= (\Sigma H_i \mathrm{Exp}(B_m^* V_i))(\Sigma H_i \mathrm{Exp}(-B_m^* V_i))/(\Sigma H_i)^2 \tag{15}$$

Putting

$$f(B_m^*) = (\Sigma H_i \mathrm{Exp}(B_m^* V_i))(\Sigma H_i \mathrm{Exp}(-B_m^* V_i)) - D (\Sigma H_i)^2 \tag{16}$$

and taking the first derivative

$$f'(B_m^*) = (\sum H_i V_i Exp(B_m^* V_i))(\sum H_i Exp(-B_m^* V_i))$$

$$- (\sum H_i Exp(B_m^* V_i))(\sum H_i V_i Exp(-B_m^* V_i)) \qquad (17)$$

the coefficient B_m^* could be evaluated by the Newtonian iteration formula

$$B_m^*(k + 1) = B_m^*(k) - f(B_m^*)/f'(B_m^*) \qquad (18)$$

with the known inhomogeneity index and the experimental chromatogram of the sample using

$$\varepsilon \geq B_m^*(k + 1) - B_m^*(k) \qquad (19)$$

as the objective function. Coefficients A_m^* are then evaluated by substituting B_m^* into Equation 13 or 14.

The third step of computation is to find $M(V_R)$ by linear regression or polynomial fitting after the crosspoint coordinates of all of the samples have been evaluated.

The forth step is to estimate the parameter ς and the spreading factor by Equation 9, 10 and 8, taking $\sigma_0^2(\bar{V})$ as an approximation of function $\sigma_0^2(V_R)$.

Experimental

Six commercial narrow MWD polystyrene standards (Applied Research Laboratories Limited, England) and five broad MWD 1,2-polybutadiene fractions were used to calibrate an ARL 950 GPC instrument with silica bead packed columns. The ARL polystyrene standards were also used to calibrate a number of home-made SEC column units packed with silica beads or styrene-divinylbenzene copolymer beads. The average molecular weights of these samples are listed in Table I. Tetrahydrofuran or toluene was used as eluents for these columns.

Table I. The Average Molecular Weight and SEC Data of Polystyrene and 1,2-Polybutadiene

Polymer		$\langle M \rangle_w 10^{-5}$	$\langle M \rangle_n 10^{-5}$	$\bar{V}$	σ_T^2
Polystyrene	A1	0.04	0.037	166.4	17.8
	A2	0.10	0.093	157.9	18.5
	A3	0.204	0.198	151.5	17.9
	A4	1.1	1.09	140.4	15.8
	A5	3.9	3.85	132.2	17.0
	A6	28.6	24.7	115.0	42.5
Polybutadiene	S1A	2.33	1.60	133.9	46.4
	S1	6.46	3.10	127.3	79.2
	S2	10.7	6.41	120.4	61.0
	S3	13.4	8.69	117.9	55.5
	S4	21.2	15.1	114.8	47.3

Results and Discussion

The mean elution volume and total variance calculated from the experimental chromatograms of polystyrene and 1,2-polybutadiene on the ARL 950 GPC instrument are also listed in Table I. The coefficients of the effective relation, coordinates of the cross-point, parameter ξ and spreading factor were computed by the scheme outlined above. The results obtained are listed in Table II and III. The effective relations and calibration

Table II. The Calculated Coefficients of the Effective Relation

Polymer		A_m^*	B_m^*	$M^*(\bar{V})10^{-5}$
Polystyrene	A1	19.19	0.0657	0.038
	A2	19.42	0.0649	0.096
	A3	16.08	0.0407	0.201
	A4	14.98	0.0241	1.10
	A5	16.51	0.0276	3.88
	A6	21.61	0.0593	26.6
Polybutadiene	S1A	24.29	0.0906	1.92
	S1	25.53	0.0984	4.49
	S2	24.76	0.0925	8.30
	S3	24.41	0.0892	10.8
	S4	24.18	0.0853	17.8

Table III. The Calculated Results of the Parameter ξ and the Spreading Factor

Polymer		ξ		σ_o^2	
		S	I	S	I
Polystyrene	A1	0.500	0.505	13.4	13.3
	A2	0.494	0.489	14.0	14.1
	A3	0.310	0.300	16.2	16.3
	A4	0.183	0.185	15.3	15.3
	A5	0.210	0.233	16.3	16.2
	A6	0.450	0.444	33.9	34.1
Polybutadiene	S1A	0.816	0.814	15.4	15.5
	S1	0.886	0.893	17.0	16.1
	S2	0.833	0.829	18.6	19.1
	S3	0.804	0.799	19.6	20.1
	S4	0.769	0.775	19.3	18.9

S: from slope,   I: from intercept.

relation for polystyrene and 1,2-polybutadiene are shown in
Figure 1 and 2 respectively. The coefficients of the calibration
relation obtained by linear regression are A_m = 30.30, 27.05 and
B_m = 0.1316, 0.1110 for polystyrene and 1,2-polybutadiene respectively. The variation of spreading factor with elution volume
derived from polystyrene and 1,2-polybutadiene are coincident and
in accord with that obtained by the method of coupling GPC with
LALLS for the same column (12) as shown in Figure 3.
   The uncertainty of the calculated spreading factor $\Delta(\sigma_0^2)$
depends upon the accuracy of the inhomogeneity index of the sample
and that of the total variance of experimental chromatogram. It
may be expressed as

$$\Delta(\sigma_0^2) = (\partial\sigma_0^2 / \partial\sigma_T^2)\Delta(\sigma_T^2) - (\partial\sigma_0^2/\partial D)\Delta D. \qquad (20)$$

If the experimental chromatogram is Gaussian, the spreading factor
could be represented by

$$\sigma_0^2 = \sigma_T^2 - \ln D / B_m^2 \qquad (21)$$

Substituting its partial derivatives into Equation 20 we have

$$\Delta(\sigma_0^2) = \Delta(\sigma_T^2) - (1 / B_m^2) (\Delta D / D) \qquad (22)$$

Thus the absolute error of the calculated spreading factor depends
upon the absolute uncertainty of total variance, the slope of the
calibration curve and the relative uncertainty of the inhomogeneity index $\Delta D/D$ of the sample. Equation 22 was verified by
arbitrarily changing the inhomogeneity index of polystyrene
standards, recalculating the spreading factor with the same compu-
ting program and plotting the deviation $\Delta(\sigma_0^2)$ versus $\Delta D/D$ as
shown in Figure 4. Therefore, if the standard sample was well
characterized, the error of the calculated spreading factor is
mainly caused by the uncertainty of the total variance. It is
shown in Figure 3 that the spreading factor derived from
polystyrene sample A6 is much larger than others'. It is
probably caused by the larger total variance due to the greater
extent of adsorption of high molecular weight polystyrene on the
porous silica gel.
   It is interesting to examine the effect of the molecular
weight of polymer and the role of the packing material on the
spreading factor. The ARL polystyrene standards were used to
determine the spreading factor of a number of SEC units packed
with styrene-divinylbenzene copolymer gel (SDG) or silica gel (SG)
beads using tetrahydrofuran or toluene as eluent by the present
method. The data cannot be compared directly because the volumes
of the siphon tubes and columns of these SEC units are different.
But if the relative value of spreading factor i.e. the ratio of
the spreading factor of the standards to that with lowest
molecular weight (A1) is considered, some interesting features
could be realized as shown in Figure 5 in which the relative
spreading factor is plotted as a function of molecular weight.
The molecular weight dependency of the spreading factor, in other
words the restricted diffusion of the macromolecule in the pore
is much pronounced for styrene-divinylbenzene copolymer gel.

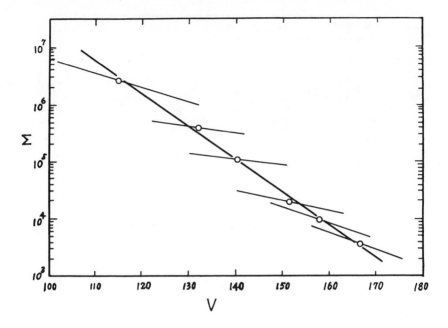

Figure 1. The calibration relation M(V_R) and effective
relations M*(V) of narrow MWD polystyrene standards.

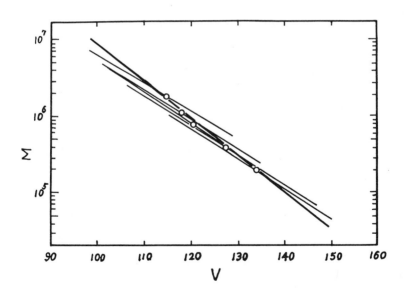

Figure 2. The calibration relation M(V_R) and effective
relations M*(V) of broad MWD 1,2-polybutadiene fractions.

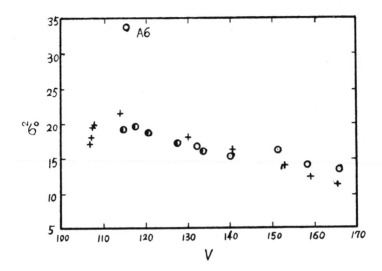

Figure 3. Variation of the spreading factor with the elution volume. Key: o, PS; ◑, PB, -, GPC-LALLS, PS.

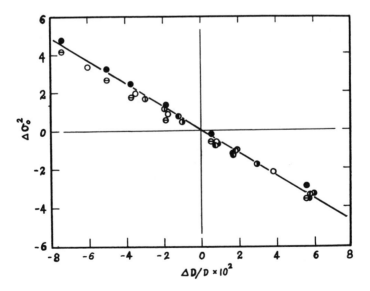

Figure 4. Dependency of the uncertainty of spreading factor on the relative error of inhomogeneity index. Key: o, Al; ◑, A2; ◐, A3; ⊙, A4; ⊖, A5; and ●, A6.

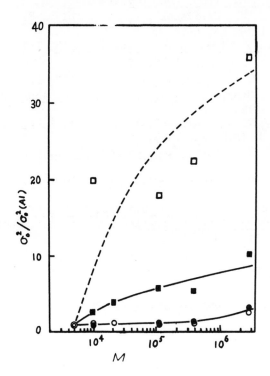

Figure 5. Variation of relative spreading factor with the molecular weight of polystyrene standards. Key: o ● SG, SEC unit 1 and 2; □ ■ SDG, SEC unit 3 and 4.

It is believed that the surface structure of the porous packing material plays an important role. The presence of the free chain ends of styrene-divinylbenzene copolymer may prevent the movement of the macromolecules in the pore.

Literature Cited

1. Balke, S. T.; Hamielec, A. E.; Leclair, B. P.; Pearce, S. L. Ind. Eng. Chem., Prod. Res. Dev. 1969, 8, 54.
2. Cardenas, J. N.; O'Driscoll, K. F. J. Polym. Sci., Polym. Lett. Ed. 1975, 13, 657.
3. Loy, B. R. J. Polym. Sci., Polym. Chem. Ed. 1976, 14, 2321.
4. Szewczyk, P. Polymer, 1976, 17, 90.
5. McCrackin, F. L. J. Appl. Polym. Sci. 1977, 21, 191.
6. Yau, W. W.; Stoklosa, H. J.; Bly, D. D. J. Appl. Polym. Sci. 1977, 21, 1911.
7. Vrijbergen, R. R.; Soeteman, A. A.; Smit, L. A. M. J. Appl. Polym. Sci. 1978, 22, 1267.
8. Chaplin, R. P.; Ching, W. J. Macromol. Sci., Chem. 1980, A14, 257.
9. Malawer, E. G. J. Polym. Sci., Polym. Phys. Ed. 1980, 18, 2303.
10. Cheng, R. S. Proceedings of China-U. S. Bilateral Symposium on Polymer Chemistry and Physics 1979, p.43 ; Gaofenzi Tongxun 1981, 123.
11. Cheng, R. S. J. Liq. Chromatogr. , in press.
12. He, Z. D.; Zhang, X. C.; Cheng, R. S. J. Liq. Chromatogr. 1982, 5, 1209.
13. Tung, L. H. J. Appl. Polym. Sci. 1966, 10. 375.

RECEIVED September 12, 1983

Evaluation of Füzes Statistical Methods

For Testing Identity of Size Exclusion Chromatography Molecular Weight Distributions of Polymers

SADAO MORI

Department of Industrial Chemistry, Faculty of Engineering, Mie University, Tsu, Mie 514, Japan

The sequential U test proposed by L. Füzes could differentiate two polymers whose molecular weight averages are identical within the experimental errors. Parallel measurements of SEC chromatograms of the two polymers were performed in series and the distinguised points (DPs), which are defined as the elution volumes at 10, 30, 50, 70, and 90 % of the each integral chromatogram, were calculated. By the statistical treatments of the DP values, identity of molecular weight distributions (MWDs) of the two polymers was established with more than four pairs of parallel measurements, and the disagreement of MWDs with two to four pairs of runs. However, this statistical treatment could not detect small differences in shapes of the both chromatograms.

Polymer samples of same species can be confirmed in their identity by the agreement with the respective values of both the molecular weight average and the molecular weight distribution (MWD). These values are to be measured by size exclusion chromatography (SEC). In the determination of SEC, we often experience the conflicts that polymer samples having the identical molecular weight averages, within the experimental errors, show different SEC chromatograms or vice versa. It is very important to know if the observed differences between MWDs or between the molecular weight averages are due to real deviations or to experimental errors. The identity of molecular weight averages can be tested by the t-test by determining these values repeatedly and by knowing the standard deviation. However, another statistical treatment must be required in the case of MWDs in order to judge the difference to be due to the experimental variations or the real MWD.

Recently, L. Fuzes reported the method of "distinguished points (DPs)" for comparing the SEC chromatograms of two or more polymer samples (1). The sequential U and t tests were suggested in order

0097–6156/84/0245–0135$06.00/0
© 1984 American Chemical Society

to indicate the significant deviation or the agreement of the DP
values of SEC chromatograms of two polymer samples. In this
report, the validity of this statistical method was tested by
using several polystyrene mixtures of known broad and narrow MWDs.
Whether the difference in the two MWDs of polystyrenes having the
similar molecular weight can be regarded as significant or not was
tested using several pairs of test samples, one is polystyrene NBS
706 and the other the mixture of polystyrene NBS 706 and another
polystyrene having different molecular weights than NBS 706.

Calculation

Basic parameters for the comparison of two polymers, A and B, are

$$h_o = -h_1' = -\frac{2\sigma^2}{\delta} \ln \left(\frac{1 - 0.5\alpha}{\beta}\right) \tag{1}$$

and

$$h_1 = -h_o' = \frac{2\sigma^2}{\delta} \ln \left(\frac{1 - \beta}{0.5\alpha}\right) \tag{2}$$

where α is the error of type I, β the error of type II, δ the
least difference of the elution volume in this case one wants to
detect, σ the standard deviation of the DP values which are the
elution volumes at 10, 30, 50, 70, and 90 % of the integral curves
of the chromatograms in the order of increasing elution volume.
     The term ΔT_{ij} is defined as

$$\Delta T_{ij} = T_{Aij} - T_{Bij} \tag{3}$$

where T_{Aij} and T_{Bij} are the DP values of each chromatogram, the
index i defines the number of parallel measurements (1, 2, -----,
n), and the index j identifies the per cent of the integral curve
of each chromatogram (10, 30, 50, 70, and 90 %). After every
parallel run of polymers A and B, compute ΔT_{ij} for each i and j
and summarize the ΔT_{ij} values for i in the case of each j

$$\sum_{i=1}^{n} \Delta T_{ij} \tag{4}$$

and plot this value at each n. The broken lines on Figure 2 are
defined as T_1 (= h_1 + Sn), T_o (= h_o + Sn), T_1' (= h_1' - Sn), and
T_o' (= h_o' - Sn) in the order from the top to the bottom, where
$S = \delta/2$ and n is the number of parallel measurements.

Experimental

SEC measurements were performed on a Jasco TRIROTAR high-performance liquid chromatograph with a Model SE-11 differential refractometer. Two Shodex A80M high-performance SEC columns (50 cm x 8
mm i.d.) packed with a mixture of polystyrene gels of nominal
exclusion limits of 10^5, 10^6, 10^7, and about 10^8 molecular weights
as polystyrene were used and thermostated at 25 °C in an air oven,
Model TU-100. The data were evaluated automatically by using a
Sord micro-computor Model 220 to which the out-put of the detector
was connected via an A/D converter.

Tetrahydrofuran was used as the mobile phase. The flow rate
of the pump dial was adjusted to 1.0 mL/min and sample concentration was 0.2 % (w/v). A 0.25-mL loop was used to inject these
sample solutions. Sample polymers were standard polystyrene NBS
706 ($\overline{M}_w$ = 2.71 x 10^5, $\overline{M}_n$ = 1.30 x 10^5 measured at our laboratory),
commercial polystyrene ESBRITE ($\overline{M}_w$ = 2.26 x 10^5, $\overline{M}_n$ = 1.08 x 10^5)
and two narrow MWD polystyrenes, PS 411000 and PS 200000 (molecu-
lar weights are 411,000 and 200,000, respectively).

Results and Discussion

First, we estimated the parameters α to be 0.01 and β to be 0.05.
For the estimation of the standard deviation of the DP values,
twenty chromatograms of NBS 706 polystyrene were measured and
elution volumes at the distinguished per cent points of the
integral curve of each chromatogram were calculated. Then, the
value σ was obtained to be 0.042 mL. The value δ was estimated
to be 0.1 mL, which corresponds to 0.3 % of the elution volume at
the center of the calibration curve of this SEC system and 5 %
difference of molecular weight.

The sequential U test was performed by using several pairs of
polymer samples. First example is shown in Figure 1. The sample
mixture is a combination of NBS 706 (98.5 %) and PS 200000 (1.5 %).
Two normalized chromatograms, NBS 706 (A) and the mixture (B), are
nearly the same and molecular weight averages (the mean from three
determinations) calculated are almost identical. The results of
the sequential U test is shown in Figure 2. After the fourth pair
of runs, all the values of $\Sigma\Delta T_{ij}$ were found to be located in the
area A = B, and it could be stated with a risk of 5 % that the two
polymer samples had the same MWDs. Normalized chromatogram of a
similar mixture of NBS 706 (97 %) and PS 200000 (3 %) is shown
with that of NBS 706 in Figure 3 and the sequential U test is
shown in Figure 4. In this example, the value of $\Sigma\Delta T_{ij}$ at j = 10
% exceeded the critical value after the fifth pair of parallel
measurements. It could be stated with a risk of 5 % that the MWD
of the mixture was not the same to that of NBS 706 though they had
the almost identical molecular weight averages. Since the

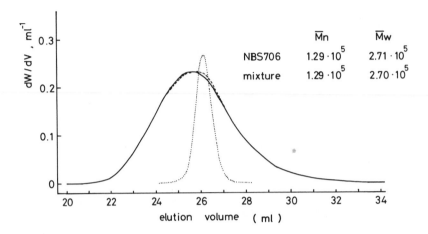

Figure 1.   Normalized SEC chromatograms of NBS 706 (————),
PS 200000 (·········), and the mixture of NBS 706 (98.5 %) and
PS 200000 (1.5 %) (————).

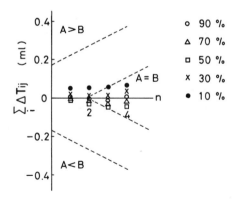

Figure 2.   Sequential U test for (A) NBS 706 and (B) the
mixture of NBS 706 (98.5 %) and PS 200000 (1.5 %).

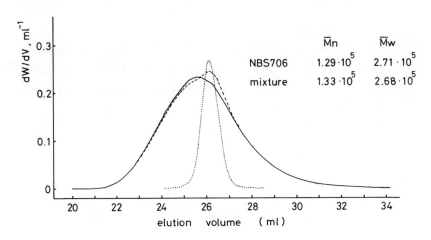

Figure 3. Normalized SEC chromatograms of NBS 706 (———),
PS 200000 (·······), and the mixture of NBS 706 (97 %) and
PS 200000 (3 %) (————).

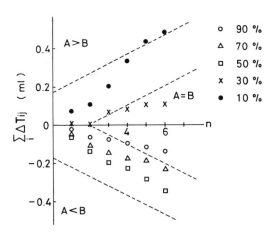

Figure 4. Sequential U test for (A) NBS 706 and (B) the
mixture of NBS 706 (97 %) and PS 200000 (3 %).

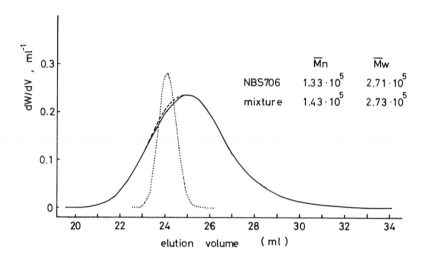

	$\overline{M}_n$	$\overline{M}_w$
NBS706	$1.33 \cdot 10^5$	$2.71 \cdot 10^5$
mixture	$1.43 \cdot 10^5$	$2.73 \cdot 10^5$

Figure 5.   Normalized SEC chromatograms of NBS 706 (———),
PS 411000 (------), and the mixture of NBS 706 (98.5 %) and
PS 411000 (1.5 %) (-- -- --).

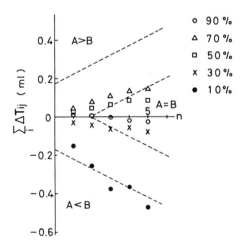

Figure 6.   Sequential U test for (A) NBS 706 and (B) the
mixture of NBS 706 (98.5 %) and PS 411000 (1.5 %).

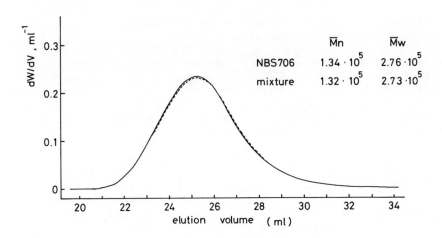

Figure 7. Normalized SEC chromatograms of NBS 706 (————)
and the mixture of NBS 706 (95 %) and ESBRITE (5 %) (————).

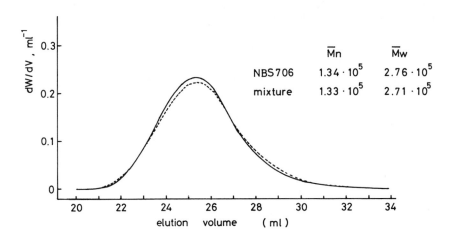

Figure 8. Normalized SEC chromatograms of NBS 706 (————)
and the mixture of NBS 706 (90 %) and ESBRITE (10 %)(————).

distinguished point at j = 10 % was located in the area A > B, it means that the MWD of the mixture was shifted to the higher molecular weights at the high molecular weight part. However, this test method could not detect the abnormality of the shape of the chromatogram at around center.

Figure 5 is the case of a mixture NBS 706 (98.5 %) and PS 411000 (1.5 %). The value of $\overline{M}_n$ of the mixture is about 8 % higher than that of NBS 706, but both values of $\overline{M}_w$ are nearly the same. Both normalized chromatograms have similar shapes as in the case of Figure 1. However, the sequential U test (Figure 6) revealed after the third pair of parallel measurements that the MWD of the mixture is different from that of NBS 706.

Figures 7 and 8 show the normalized chromatograms of NBS 706 and of the mixture of NBS 706 and ESBRITE. Molecular weight averages of each pair of runs can be regarded as identical within the experimental errors. The chromatograms in Figure 7 were judged to be identical by the sequential U test of four pairs of parallel measurements. Only two pairs of runs were necessary for the decision of disagreement in the case of Figure 8.

In conclusion, the sequential U test is useful for the judgement of identity between MWDs of a pair of polymer samples whose molecular weight averages are identical within the experimental errors. Identity of MWDs of the two polymer samples was established with more than four pairs of parallel measurements, and the disagreement of MWDs with two to four pairs of parallel measurements. Though this statistical treatment is useful for the identification or differentiation of the MWDs of the pair of polymers, it can not detect small differences in shapes of the both chromatograms.

Literature Cited

1. Füzes, L. J. Appl. Polym. Sci. 1979, 24, 405-416.

RECEIVED September 27, 1983

OPERATIONAL VARIABLES, COLUMN TECHNOLOGY, AND OLIGOMERS

A New Family of Organic Polymer-Based High-Efficiency Gel Permeation Chromatography Columns

HERMAN S. SCHULTZ, PETER G. ALDEN, and JURIS L. EKMANIS

Waters Associates, Milford, MA 01757

The ULTRASTYRAGEL family of columns for separation by molecular size was studied using calibration curves and Probe Mixtures. The Probe Mixtures consisted of combinations of small molecules, polystyrene oligomers and high molecular weight polymer standards. Two experimental mixed pore size columns with very broad pore size distribution were also evaluated. Columns (30 cm long) were evaluated for their ability to resolve the Probe Mixtures using various combinations of one to four columns. The Probe Mixtures serve as qualitative but visually very apparent indicators of resolving power. Use of such Probe Mixtures can facilitate understanding of the interaction among amount of pores, distribution of pore sizes, number of plates, and resolving power. This in turn leads to optimum utilization of combinations of the columns. Shorter analysis times can then be attained utilizing sets of one or two columns of the proper pore size.

In 1964 Moore and Hendrickson (1,2) introduced the technique of "Gel Permeation Chromatography" (GPC) for determining molecular weight distributions of polymer samples. Moore's work introduced the use of chromatographic column packings consisting of then considered small porous spherical organic polymer particles (37-75u). These particles were made from highly crosslinked copolymers of styrenes and divinyl benzenes. They became available as a family of columns under the name STYRAGEL. Subsequently, much more efficient families (3,4) of columns became available as particle sizes were reduced. The columns are generally appropriate for resolution of oligomers through very high molecular weight polymers that are soluble in organic solvents.

The mechanism of separation was by molecular volume or

0097–6156/84/0245–0145$07.25/0

size. This was equated with molecular weight after calibration with very narrow molecular weight distribution standards. These standards were defined by primary molecular weight measurements such as light scattering, ultracentrifugation, osmometry, etc.. Historically, and even today, polystyrene standards are the most readily available. They are defined with least ambiguity. The subject packings exhibit little or no adsorption (non-size) effects when used to resolve compounds and polymers in appropriate mobile phases. It was noted early in the history of the subject that columns of appropriate pore size could even be used to resolve mixtures of small organic molecules (5,6,7).

Packings based on silicas for separation by molecular size are also presently available. However, the molecular size range available is more limited and possibility of encountering adsorption effects is more likely. Such columns must be throughly evaluated for each new type of sample for which a separation and/or molecular weight distribution is desired.

One subject of this paper is the description and illustration of the chromatographic characteristics and capabilities of a new family of styrene based GPC columns designated by the name ULTRASTYRAGEL (4,8,9,10). The possibilities created by the considerable increase in efficiency leads to the need for reassessment of how to evaluate and utilize columns of different pore size ranges. This is relative to the extended banks of columns conventionally used. Much higher speed with greater resolution than hitherto possible can now be attained in a given situation. Alternatively, extraordinary resolution is possible when time is not an issue and extended banks of these columns can be used. This new family of columns is made possible by new suspension polymerization processes for small particles (11), coupled with improved insights into the relationships of particle size distributions and the art of column packing.

Carefully constructed Probe Mixtures based on small molecules and polystyrene standards are used as standardized reference points to better define the functional capabilities of individual columns and column combinations. The result using the method of Probe Mixtures to evaluate columns is better than can be attained from calibration curves alone and is especially useful in this high resolution capability situation.

Experimental

In most cases, calibration curves were determined at ambient or elevated temperatures using the Waters 150C High Temperature Gel Permeation Chromatograph which includes a sensitive refractive index detector. Otherwise, a modular system consisting of a Waters Model M6000A Solvent Delivery System, a Waters Model U6K Injector and a Waters Model 401 Refractometer, were used at ambient temperature. The mobile phase at room temperature was toluene and at $140^{\circ}C$ was 1,2,4-trichlorobenzene. Standard flow rate was 1 ml/minute. Model 401 Refractometer sensitivity was 4X or 8X. Polystyrene standards were obtained from Waters Associates, Milford, MA, and Toyo Soda Manufacturing Co., Japan. Especially great care was taken in handling standards above one million molecular weight to minimize the possibility for shear degradation. These standards were used at 0.02% concentration, prepared fresh daily, and 50-100 ul of solution was injected per column. Plates were determined using ortho dichlorobenzene and corroborated with dicyclohexyl phthalate, resulting in similar values for all columns, except with the $10^{6}A$ columns where the dicyclohexyl phthalate value was used. These markers were injected as 3-5% (w/v) solutions (10 ul). Both the tangent and 5 sigma methods were used to calibrate plates (12) and both methods were used to judge the quality of a column.

Special attention was paid to minimizing band spreading due to instrumentation since this is especially deleterious to efficiency when very high plate columns are used. A measure of band spreading was determined by measuring the volume of the band width of a 10 ul injection of 3% ortho dichlorobenzene with no column in the instrument and a minimal volume connector. The length of tubing in the system was kept as short as possible and only 0.009" I.D. tubing was used between the injector and detector. Samples were injected immediately after loading into the injector to minimize diffusion of the sample in the sample loop. The volume of the band width was calculated by the equation,

$$\text{System Band Spreading (ul)} = (W_{5\sigma})\ (F)\ (1000)/(CS)$$

where W_5 is peak width at 4.4% peak height (cm.), F is flow rate (ml/minute), CS is chart speed (cm/minute). Typical band spreading within the instrument should be 100 ul or less. Band spreading significantly greater than 100 ul indicates an instrument problem that must be corrected.

Figure 1 defines the Probe Mixtures based on small molecules through high molecular weight polystyrene standards and the concentrations and volumes used per column. The ULTRASTYRAGEL family at ambient temperatures can be used with organic solvents such as toluene, tetrahydrofuran, methylene chloride, chloroform, etc. It has also been used at elevated temperatures with appropriate solvents such as chlorinated benzenes, cresols, and dimethyl formamide for polyolefins, polyesters and other polymers requiring elevated temperatures. All figures are based on ULTRASTYRAGEL columns.

MIX 1   Benzene, ortho xylene, 50/50 by volume

MIX 2   0.5% Polystyrene Oligomer Mix "300" — 10 identified components
                                                              MW 161-1098, peak at 370

MIX 3   2.0% Bezene
         0.16% Polystyrene Oligomer Mix "300"
         0.10% Polystyrene Oligomer Mix "1000"
         0.03% 2800 MW Polystyrene Standard
         0.03% 6200 MW Polystyrene Standard

MIX 4 (0.03% of each)   MIX 5 (0.03% of each)

	Ratio Successive Components			Ratio Successive Components
--------	-----------------------------		--------	-----------------------------
MW			MW	
2,800	2.21X		422,000	2.99X
6,200	1.65X		1,260,000	4.35X
10,200	1.64X		5,480,000	
16,700	2.56X			
42,800	2.50X		MIX 6 (0.03% of each)
107,000	1.74X			
186,000	2.27X		1,260,000	6.68X
422,000			8,420,000	

Injection Volume: Mix 1: 0.50 µl / column, neat
                     Mix 2: 30µl, 0.5% solution/column
                     All Other Mixtures: 50 µl/column

Figure 1.   Definition of small molecules and polystyrene probe mixtures 1 to 6; injection volumes and concentrations.

Results and Discussion

Definitions of Columns Using Calibration Curves

ULTRASTYRAGEL columns (Figure 2) are available in six pore sizes ($100A^O$ to 10^6A^O designation) and are the same with respect to pore size distribution as larger particle size STYRAGEL and uSTYRAGEL columns. The total molecular weight range is from approximately 50 (small molecules) to over ten million molecular weight based on polystyrene standards. This is the approximate upper limit for valid use of such standards (13). The second column in Figure 2 tabulates a conservative estimate of the optimum molecular weight range for each column based on interpretation of calibration curves developed using small molecules and polystyrene standards. The third column in Figure 2 indicates the often broader utility range as shown by the use of standard Probe Mixtures. The fourth column in Figure 2 lists minimum column efficiencies in terms of plates. Most columns significantly exceed these minimum values.

The log molecular weight vs. elution volume calibration curves of the six individual 30 cm long columns is presented in Figure 3. The highest molecular weight polystyrene standard used was 8.4 million. For the 10^6A^O column, it is apparent that the exclusion limit has not yet been reached. Figure 4 illustrates the same for banks of three columns consisting of 10^4A^O, 10^5A^O and 10^6A^O, or $100A^O$, $500A^O$ and 10^3A^O. Figure 5 presents calibration curves for a four column bank consisting of 10^3A^O through 10^6A^O and, for comparison, two sets of two columns consisting of 10^3A^O plus 10^5A^O and 10^4A^O plus 10^6A^O. The middle curve for the bank of four columns, plotted using a half scale for comparison purposes, is essentially "linear" for most of its length. This historically has been considered desirable for calibration purposes although moderately sloping curves today can be handled readily by the use of computer based methodology.

With the subject columns, the augmented resolving power, due to high plates, of a relatively smaller amount of pores in a given pore size range   becomes useful for calibration purposes in non-linear portions of curves. The 10^4A^O plus 10^6A^O column combination in Figure 5 is a good example of this. It is relatively deficient in pore amount at the lower molecular weight end but has greater capability than the comparable 10^3A^O plus 10^5A^O combination at the high molecular weight end to an undetermined degree beyond the highest molecular weight standard. This is indicated by the use of Probe Mixtures to be discussed and confirmed by mercury porosimetry measurements of pore size. The 10^3A^O plus 10^5A^O combination in Figure 5 is close to, but not quite, linear. However, it has been used to obtain approximate molecular weight distributions.

COLUMN	OPTIMUM MW RANGE (Judged from calibration curves)	FUNCTIONAL MW RANGE (Judged from probe mixtures)	MINIMUM COLUMN EFFICIENCIES, N_{tan}
100 Å	50 - 1,500	50 - 1,500	10,000 ppf
500 Å	100 - 10,000	100 - 15,000	14,000 ppf
10^3 Å	200 - 30,000	200 - 40,000	14,000 ppf
10^4 Å	5,000 - 600,000	3,000 - 1,000,000	14,000 ppf
10^5 Å	50,000 - 4,000,000	30,000 - 8,000,000	14,000 ppf
10^6 Å	200,000 - $\geq$ 10,000,000	200,000 - $\geq$ 10,000,000	14,000 ppf

MIXED PORE

MP-35	1,000 - 4,000,000	500 - 8,000,000	14,000 ppf

("D" Type, see text)

Figure 2. Ultrastyragel GPC column specifications based on polystyrene standards, toluene as mobile phase (1ml/min).

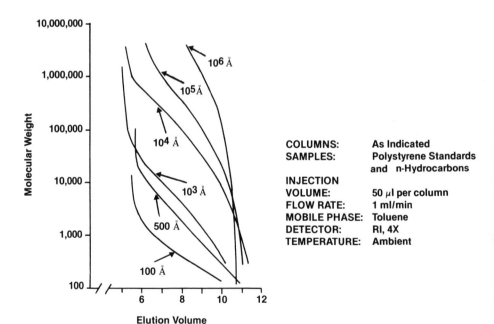

COLUMNS: As Indicated
SAMPLES: Polystyrene Standards and n-Hydrocarbons

INJECTION
VOLUME: 50 µl per column
FLOW RATE: 1 ml/min
MOBILE PHASE: Toluene
DETECTOR: RI, 4X
TEMPERATURE: Ambient

Figure 3. Individual Ultrastyragel column calibration curves.

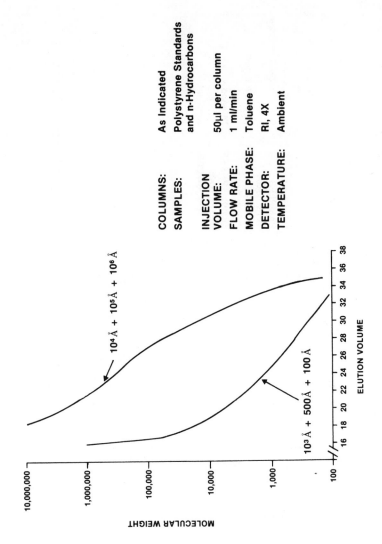

Figure 4. Calibration curves — three column sets.

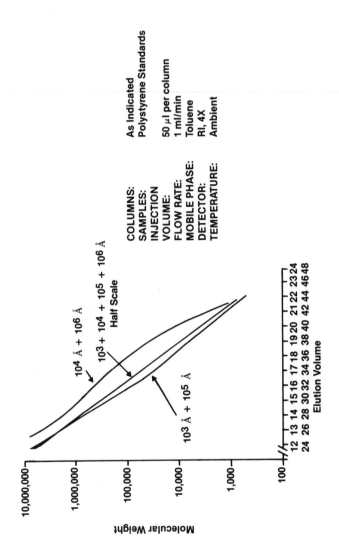

Figure 5. Calibration curves – comparison selected two column sets with four column set.

Figure 6 shows calibration curves for three other two column combinations, each representing 30,000 to 40,000 plates per set. The 10^3A^o plus 10^6A^o curve can be interpreted to show a deficiency in relative pore population in the range equivalent to about 50,000 to 600,000 molecular weight. The other two, properly calibrated, can conceivably be used for determination of molecular weight distributions. However, utility for resolution of specific polymodal mixtures is too difficult to assess from calibration curve alone. How much curvature of a calibration curve translates into utility or non-utility? Calibration curves indicating pore size populations all have the same shape for given column combinations whether the plate count level is 5000 plates or 20,000 plates or 80,000 plates.

Calibration curves of two column banks each consisting of two experimental mixed pore columns are presented in Figure 7. The calibration curves were determined at 140°C with trichlorobenzene as mobile phase. Each individual column in a set has exactly the same pore size distribution so that each can be used individually if the resolving capability is sufficient for a specific situation. The calibration curves for the "NW" type and the "D" type column banks should be compared respectively to the 10^4A^o plus 10^6A^o and 10^3A^o plus 10^5A^o banks in Figure 5. The comparisons indicate that it is possible to attain a very wide range of capabilities for screening and many quality control purposes with a single 30cm column. The single column is operated at 1 to 1.5 ml/minute, and between exclusion times of 4-6 minutes and total permeation time of 8-12 minutes. It should be remembered that all events take place within one volume of pores of a column or bank of columns. The "D" Type (or MP-35) mixed pore column is linear for a major portion of its length as shown in Figure 7.

The availability of calibration curves for individual columns and various combination banks of columns affords no more than a general insight, based on pore distribution, into the performance of the very high resolving power columns. The use of carefully constructed standard Probe Mixtures will now be discussed to evaluate in more detail the capability of different column combinations.

Use of Probe Mixtures to Define Columns and Their Performance

Probe Mixtures serve as visually very apparent indicators of resolving power. The mixtures were constructed and standardized to cover the molecular weight range from small organic molecules through high molecular weight polymodal mixtures. Figure 1 defines Probe Mixtures 1 to 6. The components of a mix were chosen so that there would not likely be baseline resolution unless there was a very favorable interplay between a high plate value and the amount of pores in a given size range, resulting

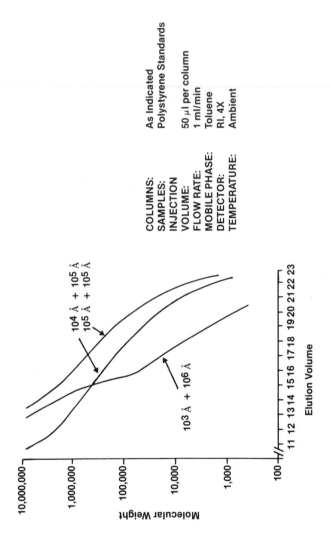

Figure 6. Calibration curves – selected two column sets.

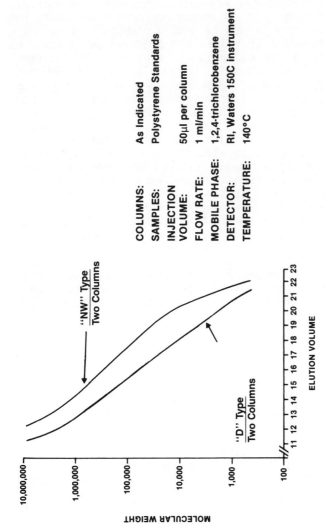

Figure 7. Calibration curves – two column sets experimental mixed pore columns.

in the needed resolving power. In Figure 1, the tabulations for Mixes 4, 5 and 6 indicate the small values chosen for the ratios of molecular weights of the adjacent standards.

The critical operational assumption that makes it possible to draw conclusions in a given comparison situation about the effect of plate and pore amount is that a constant volume and a constant absolute amount of solute was injected per column to normalize comparisons. If pore amount per column is constant, then increase in resolution with several columns of the same kind in series is due only to the increased amount of plates. Conversely, if plates of a column bank are the same, then differences in resolution are due to differences in the amount of pores of appropriate size. Also, all the other appropriate operating parameters are constant for each comparison. The following group of comparisons will illustrate different issues involving the interplay of pores, plates, and resolving power. The times on the figures are maximum values for total permeation volumes at a flow rate of 1 ml/min.

Small molecule Mix 1 (MW=78, 106) is used in Figure 8 with sets of three, two, or one 10^3A° columns. This figure illustrates that a large amount of plates makes up for insufficient pores to attain resolution. The resolution is functional evidence for the existence of sufficient pores of appropriate size. The previous generation of lower efficiency (i.e. 5000 plates per column) 10^3A° columns were never considered to have resolving power in this molecular weight range. The same probe is used in Figure 9. Two $100A^\circ$ and two $500A^\circ$ column banks are compared with the $500A^\circ$ columns having twice as many plates. The $100A^\circ$ bank having more pores in the appropriate range, resulted in approximately the same degree of resolution.

The same points are illustrated in Figure 10 using Mix 2, a polystyrene oligomer mix with components in the 161 to 1,098 range and highest population peak at 370. The bottom row across the figure is a comparison of the same amount of plates. The $100A^\circ$ set gives the best results at the same plate level. In the top row where all sets contain three columns, the $500A^\circ$ bank having twice as many plates is comparable to the $100A^\circ$ bank. It should be noted that the 10^3A° bank still has usefulness in this range for screening purposes due to high plates.

The remainder of this paper illustrates that families of reference chromatograms can be developed to determine if a given column combination is the best one to be used with an unknown polymer or polymer mixture, the families of chromatograms being based on the standard Probe Mixtures, one to four column combinations and different plate levels. Extended banks of lower efficiency columns would be required to attain the same degree of resolution of even one of these ULTRASTYRAGEL columns.

Figure 11 indicates the performance of three and one column banks of 10^3A°, 10^4A° or 10^5A° columns using polymodal Probe Mix

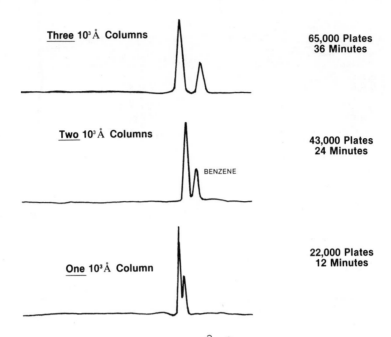

Three 10³Å Columns — 65,000 Plates / 36 Minutes

Two 10³Å Columns — 43,000 Plates / 24 Minutes

BENZENE

One 10³Å Column — 22,000 Plates / 12 Minutes

Figure 8. Comparison 3, 2, 1 10^3 A^o column sets illustration larger amount of plates makes up for insufficient pore amount; Probe Mix 1, benzene and ortho xylene.

Two 100Å Columns — 24,000 Plates / 22 Minutes

Two 500Å Columns — 50,000 Plates / 23 Minutes

Figure 9. Comparison $100A^o$ and $500A^o$ two column sets; plates versus pore amount; Probe Mix 1.

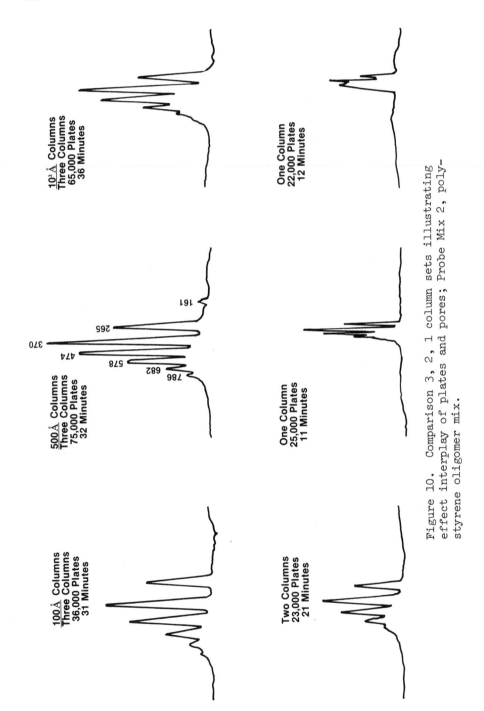

Figure 10. Comparison 3, 2, 1 column sets illustrating effect interplay of plates and pores; Probe Mix 2, polystyrene oligomer mix.

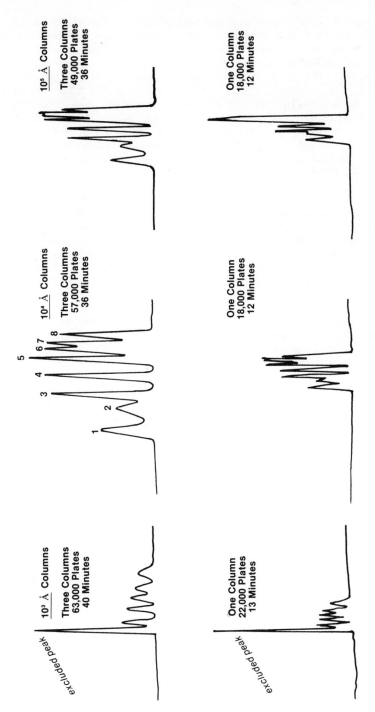

Figure 11. Comparison 3 and 1 column sets of same pore size, illustrating effect interplay of plates and pores; Probe Mix 4. MW 1 to 8 is 422,000, 186,000, 107,000, 42,000, 16,700, 10,300, 6,200, and 2,800.

4 consisting of eight polystyrene standards. The diagnostic patterns for optimum use range of each column type can be seen as a function of pore size and distribution at comparable plate levels. The $10^3A°$ columns resolve well in the 2,800-42,000 standard range. The $10^4A°$ columns are optimum in 16,700 through 422,000 standard range. The $10^5A°$ columns function to a slightly less degree in the optimum range for the $10^4A°$ columns but poorly in the $10^3A°$ range. However, any activity at all in the $10^3A°$ range would be unexpected with previously available lower plate level $10^5A°$ columns. The optimum combination for Mix 4 would consist of $10^3A°$ and $10^4A°$ columns rather than the standard mixed banks used conventionally.

A similar comparison is shown in Figure 12 using Mix 5 in the 422,000 to 5.48 million range. It can be interpreted in a similar manner to draw conclusions about effectiveness versus pore size and increased resolution due to increased plates, everything else being equal. The 16 minute marker for expected exclusion volume of the three column $10^6A°$ chromatogram indicates considerable amount of pore size volume that is available for components greater than 5.48 million standard. The $10^4A°$ and $10^5A°$ columns have better performance than previously expected.

Four banks of columns are used in Figure 13 to determine the optimum three column combination for Mix 5. Each bank has the same level of total plates (approximately 50,000). The best result is with the $10^6A°$ bank.

Patterns of capabilities are developed in Figures 14 through 18 using several Probe Mixtures for two column combinations and the full $10^3A°$, $10^4A°$, $10^5A°$ and $10^6A°$ bank calibrated in Figures 5 and 6. Comparisons reveal the power of the two column combinations and better define the preliminary ranges of use obtained from the calibration curves. The capabilities of a number of two column combinations are illustrated in Figure 14 relative to a four column conventional bank in the molecular weight range of Probe Mixture 4. For example, the $10^4A°$ plus $10^5A°$ set shows considerable $10^3A°$ activity and the $10^4A°$ plus $10^6A°$ set shows less. It can therefore be concluded that the 10^3Ao column in the first set is contributing $10^3A°$ activity. In choosing between a 500A° plus $10^5A°$ set and a $10^3A°$ plus $10^5A°$ set for use in the range of the Probe Mixture, the second set has more activity (and effective pores) in the approximately 100,000 to 190,000 molecular weight range. Therefore, it would be the column set of choice.

Figures 15 through 18 should be examined as a group. Patterns of performance and comparisons are developed for Probe Mixtures 3, 4, 5 and 6 using two column sets $10^3A°$ plus $10^5A°$, $10^4A°$ plus $10^6A°$, Experimental "D" Type Mixed Pore and Experimental "NW" Type Mixed Pore. They all have activity down to a surprisingly low molecular weight range. Each figure shows a comparison using a single Probe Mixture. The Probe Mixtures

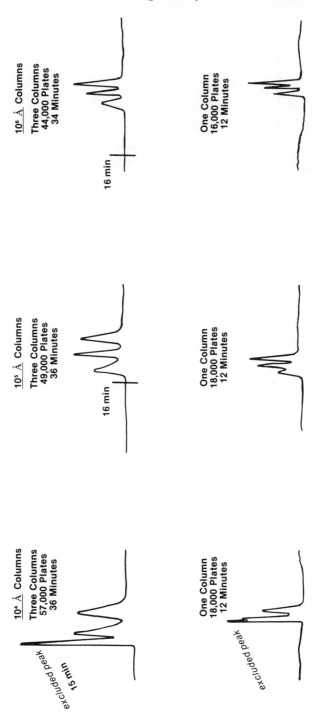

Figure 12. Comparison 3 and 1 column sets of same pore size, illustrating effect interplay of plates and pores; Probe Mix 5. MW is 5,480,000, 1,260,000, and 422,000.

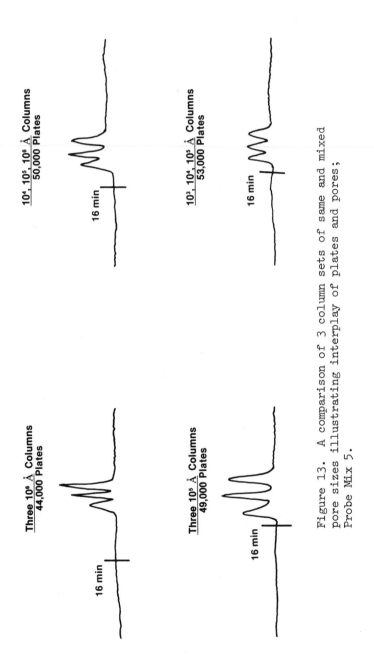

Figure 13. A comparison of 3 column sets of same and mixed pore sizes illustrating interplay of plates and pores; Probe Mix 5.

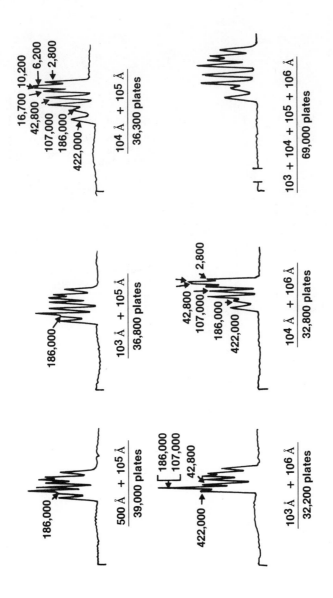

Figure 14. A comparison of capabilities of six two column sets using Probe Mix 4.

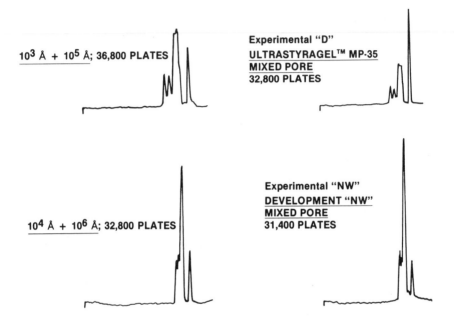

Figure 15. Patterns of capabilities; two selected individual or mixed pore column sets; Probe Mix 3.

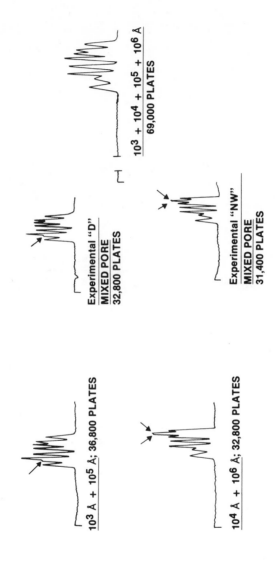

Figure 16. Patterns of capabilities; two selected individual or mixed pore column sets plus reference four column set; Probe Mix 4.

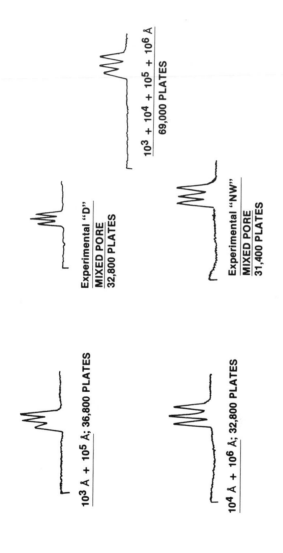

Figure 17. Patterns of capabilities; two selected individual or mixed pore column sets plus reference four column set; Probe Mix 5.

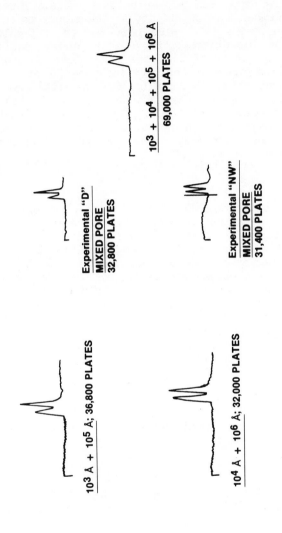

Figure 18. Patterns of capabilities; two selected individual or mixed pore column sets plus reference four column set; Probe Mix 6.

have overlapping molecular weight ranges. Together they constitute a chromatographic fingerprint for the performance of each set. Overall, the Experimental "D" Type and the $10^3 A^o$ plus $10^5 A^o$ sets have a broader molecular weight utility range than the Experimental "NW" Type and $10^4 A^o$ plus $10^6 A^o$ sets. However, the latter have somewhat more capability at the higher molecular weight pore size end. The $10^3 A^o$ plus $10^5 A^o$ set is difficult to distinguish from the Experimental "D" Type set on the basis of Probe Mixtures alone. A benefit of the Mixed Pore columns is that they combine the range of activity into one 30 centimeter very high plate count column and can be used to attain maximum speed in situations where the amount of resolution is sufficient. Another benefit of the "D" Type column is that it is "linear" over a major portion of the molecular weight range, simplifying its use for molecular weight distribution calculation purposes.

Conclusions

It is important to understand the interplay of pore amount and pore size distribution versus plates on column resolving power. This is necessary to fully utilize the performance capabilities of the new ULTRASTYRAGEL family of columns to obtain the optimum high resolution and speed appropriate for a specific use situation. Calibration curves are useful to put one into the right separation range but the use of carefully constructed standard Probe Mixtures define more specifically the performance molecular weight range of the subject columns. Stated in other terms, if one has a large amount of plates, the less pores in a given range are required for resolution, everything else being equal. "Sufficient pore amount" is defined functionally by the ability to resolve in a specific stiuation. "Sufficiency" level of pore amount is less with increasing plates. Very high efficiency columns offer the capability to resolve many mixtures of small organic molecules and polymodal polymer products without the methods development needed when separations are attempted by other mechanisms.

Literature Cited

1. Moore, J.C., J. Polym. Sci., A2 835 (1964)
2. Moore, J.C., Hendrickson, J.G., J. Polym. Sci., Part C.
    No. 8, 233 (1965)
3. Limpert, R.J., Cotter, R.L., Dark, W.A., Amer. Lab. 6
    #5, 63 (1974)
4. Schultz, H.S., Ekmanis, J.L., Tisdale, V.R., Baptiste,
    A.J., Crossman, L.W., paper presented at Pittsburgh
    Conference on Analytical Chemistry and Applied
    Spectroscopy, Atlantic City, N.J., March 8-13, 1982.
    Abstract No. 392.

5. Smith, W.B., Kollmansberger, A., J. Phys. Chem. 69 4157
(1965)
6. Hendrickson, J.G., Moore, J.C., J. Polym. Sci., A4 167
(1966)
7. Cazes, J., Gaskill, D.R., Separation Sci., 2 421 (1967)
8. Schultz, H.S., Alden, P.G., Ekmanis, J.L., paper
presented at Pittsburgh Conference on Analytical
Chemistry and Applied Spectroscopy, Atlantic City,
N.J., March 8-13, 1982. Abstract No. 393.
9. Schultz, H.S., Alden, P.G., paper presented at
Pittsburgh Conference on Analytical Chemistry and
Applied Spectroscopy, Atlantic City, N.J.,
March 7-12, 1983. Abstract No. 955.
10. Schultz, H.S., Alden, P.G., Ekmanis, J.L., paper
presented at 185th ACS National Meeting, Seattle, WA,
March 20-25, 1983. Abstract No. ORPL200; Organic
Coatings and Applied Science Proceedings, 48 945
(1983)
11. Schultz, H.S., unpublished work.
12. Yau, W.W., Kirkland, J.J., Bly, D.D., "Modern Size
Exclusion Chromatography" Wiley-Interscience, New York,
1979.
13. Slagowski, E.L., Fetters, L.J., McIntrye,
D.,Macromolecules 7 394
(1974)

RECEIVED October 13, 1983

Optimization of Resolution in Gel Permeation Chromatographic Separation of Small Molecules

F. VINCENT WARREN, JR., BRIAN A. BIDLINGMEYER, HAROLD RICHARDSON, and JURIS L. EKMANIS

Waters Associates, Milford, MA 01757

Gel permeation chromatography (GPC) has mainly been used for the determination of molecular weight distributions of polymers. The potential for using GPC for the separation of discrete small molecules has long been recognized (1-7) but practical considerations have limited this application. Until recently, the efficiency of commercially available GPC columns has been relatively low, requiring the use of sets of three or four columns to separate small molecules which have a similar effective size in solution. The expense and long analysis times associated with such a bank of columns has limited the appeal of "small molecule GPC" (SMGPC).

In 1982, the ULTRASTYRAGEL family of GPC columns was introduced (8,9). These high-efficiency columns provide a two- to three-fold increase in efficiency (plates/foot) over the closely related STYRAGEL columns which have been available since 1974. Small molecule separations which once required several STYRAGEL columns can now be performed on a single ULTRASTYRAGEL column. This advance makes SMGPC considerably more attractive as a simple and effective technique for the analysis of a variety of samples (10-12). Examples presented below will illustrate the capability of single ULTRASTYRAGEL columns in applications which are not as easily solved by other modes of HPLC (e.g. reversed phase.)

Factors Which Influence Resolution. In order to effectively apply SMGPC to separation problems, the influence of three factors on the resolution of sample components must be considered. Solvent effects play a minor role, but choice of eluent can alter selectivity in some cases. Column efficiency, as noted, has a major impact on the quality of separation. The number of peaks which can be resolved within the pore volume of a given column (i.e. peak capacity) is related to the square root of the number of theoretical plates (13). Finally, the nature of the calibration curve will influence resolution. Each

0097-6156/84/0245-0171$06.00/0

of these factors is further examined, with appropriate examples, in the discussion below.

Initially, it might seem that only one of the above three factors, the choice of eluent, could be readily adjusted by the user of commercial GPC columns. This is not the case since families of columns with varying efficiencies and pore size distributions are commercially available (14). An understanding of the other two factors will aid in selecting the most appropriate column(s) for a particular analysis and will facilitate the correct interpretation of the resulting separation. With a given eluent, improved resolution may be approached through either of two pathways: increased efficiency or a more favorable slope of the calibration curve.

The concept of "specific resolution" (Rsp) developed by Yau and coworkers (15) reflects the dual influence of efficiency and slope on resolution:

$$Rsp = 0.576/(D_2\sigma) \qquad (1)$$

D_2 is the slope of the linear portion of the calibration curve and σ is the standard deviation of the elution profile, as reflected in the peak width. According the equation 1, resolution is maximized when the product of slope and peak width (an efficiency contribution) is minimized. The interplay between these two factors must, therefore, be considered in the evaluation of available GPC columns (9,16,17).

Determination of Pore Size Distributions. The shape and range of a GPC calibration curve are, in part, a reflection of the pore size distribution (PSD) of the column packing material. A consideration of the nature of PSDs for the ULTRASTYRAGEL columns to be used in this work is therefore appropriate. The classical techniques for the measurement of PSDs are mercury porisimetry and capillary condensation. The equipment required to perform these measurements is expensive to own and maintain and the experiments are tedious. In addition, it is not clear that these methods can be effectively applied to swellable gels such as the styrene-divinylbenzene copolymer used in ULTRASTYRAGEL columns. Both of the classical techniques are applied to dry solids, but a significant portion of the pore structure of the gel is collapsed in this state. For this reason, it would be desirable to find a way to determine the PSD from measurements taken on gels in the swollen state in which they are normally used, e.g. a conventional packed GPC column.

Such a technique does, in fact, exist. In a series of papers starting in 1975, Halasz described a method for the determination of PSDs by GPC (18-25). Similar techniques have since been discussed by others (26-29). In this method polystyrene standards serve as probes of pore size. Each standard is associated with a characteristic random coil diameter in solution, and it is assumed that for each standard there will be a minimum pore diameter (∅) which allows unhindered access to that standard. An empirical equation serves to relate ∅ to the weight-average molecular weight (MW) of a polystyrene standard:

$$\emptyset \text{ [A] } 0.62(MW)^{0.59} \tag{2}$$

The elution volume for each standard is expressed according to Equation 2:

$$R = [(V_E - V_{EX})/ (V_{IN} - V_{EX})] \text{ 100\%} \tag{3}$$

where V_E is the elution volume and V_{EX} and V_{IN} are the column exclusion and inclusion volumes. For a given polystyrene standard associated with an elution volume V_E and a pore diameter ∅, R is interpreted as the percentage of the total pore volume which is formed by pores having a diameter greater than ∅. Inspection of Equation 3 reveals that R/100% = K_{GPC}, the distribution coefficient for GPC (13).

It is worth noting at this point that relatively little effort has been directed toward establishing a detailed relationship between PSDs determined by GPC (equations 2-3) and those obtained by classical methods. In the remainder of this discussion, all occurrences of the term "PSD" refer to the PSD as determined by GPC. We consider this method to yield an "effective PSD" for reasons discussed later. A discussion of the correlation between results of the GPC technique and classical methods is beyond the scope of this text.

In a previous report (30), we found that n-hydrocarbon standards are useful to extend the range of (small) pore sizes which may be probed. Polystyrene-equivalent molecular weights (MW_p) are assigned to each hydrocarbon using the empirical relationship (31):

$$MW_p = 2.3 \text{ MW} \tag{4}$$

which was derived from the analysis of GPC calibration curves.

GPC calibration data (V_E, log MW) are transformed according to equations 2 and 3 and the resulting (log ∅, R) values are plotted. Halasz has demonstrated (19) that this plot represents the cumulative PSD. The point-by-point derivative of the cumulative PSD is the "differential" PSD, which gives a rough outline of the PSD for the column packing material. A better way of obtaining the PSD is by fitting data from the cumulative PSD to a Gaussian distribution by a plot on probability paper.

Prediction of Calibration Curves. If the PSDs of individual columns can be accurately represented by Gaussian distributions, then it should be possible to predict the PSD and cumulative PSD for any combination of the individual columns. The overall PSD for a set of columns is obtained by summing the Gaussian PSDs for the individuals. Before summing, the area under each individual Gaussian is made proportional to the pore volume ($V_{IN}-V_{EX}$) of the related column. After summing, the overall PSD is integrated to obtain the cumulative PSD for the column set.

The cumulative PSD will accurately predict the calibration curve for the column set if a simple model for retention applies (13):

$$V_E = V_{EX} + \int_{\bar{r}}^{\infty} P(r)\,dr \qquad (5)$$

where V_E and V_{EX} are defined as before and $\bar{r}$ is the pore radius which is just large enough to permit unhindered access to the solute under consideration. The integral is over the pore volume consisting of all pores having a radius greater than or equal to $\bar{r}$. This model states that retention in GPC is simply governed by the fraction of the pore volume which is accessible to the given solute. If Equation 5 is correct, then the cumulative PSD contains sufficient information for prediction of the calibration curve, as indicated in Equation 6:

$$V_E = V_{EX} + \frac{V_p}{100} \int_{\bar{p}}^{\infty} R(p)\,dp \qquad (6)$$

Here V_p is the column pore volume ($V_{IN} - V_{EX}$) and $R(p)$ is the cumulative PSD, where $p = \log \emptyset$ and $\bar{p}$ is defined analogously to $\bar{r}$ above. In actual practice, the required integration may be performed graphically or approximated by computer using a simple integration algorithm.

Previous efforts (32-34) to apply the simple model of equations 5 and 6 have not yielded accurate predictions of calibration curves. The important difference between those efforts and the present work is in the source of the PSD information. Classical methods (porisimetry, capillary condensation) have been used before, rather than the GPC method described above. When classical methods are used a more complex model is required for prediction (13):

$$V_E = V_{EX} + \int_{r}^{\infty} K_{GPC}(\bar{r},r)\,P(r)\,dr \qquad (7)$$

This convolution equation takes into account the fact that K_{GPC}, the distribution coefficient, is a function of both the effective solute radius ($\bar{r}$) and the pore radius (r). Even if all the pores of a gel have the same diameter, solutes of different size can be separated on the basis of their different degrees of penetration into the pores. Thus a PSD measured by classical methods will behave as a broader PSD in practice, giving a calibration curve which could not be predicted on the basis of the classical PSD alone.

For PSDs measured by GPC, we expect a greater degree of success with the simple model for retention (eq. 5). Halasz noted that the PSDs he measured were always broader than corresponding PSDs from porisimetry and capillary condensation. This is in keeping with the convolution model (eq. 7) and indicates that the PSDs measured by GPC already contain the convolution between K_{GPC} and the classical PSD. If this is the case, then the "effective PSDs" provided by the GPC method should be useful for the direct prediction of calibration curves.

Experimental Section

<u>Chromatographic System.</u> The isocratic liquid chromatograph used was a Waters Associates (Milford, MA) Model 244 ALC which included a Model 6000A Solvent Delivery System, a Model 401 Differential Refractometer and a Model 440 Absorbance Detector operating at 254 nm and was fitted with a WISP automatic injector. The analog outputs of the UV absorbance detector or differential refractometer were recorded with a Model 730 Data Module (printer, plotter, integrator)(Waters). Eluent flow rate was 1.0 ml/min unless otherwise noted.

ULTRASTYRAGEL columns of 100A, 500A, and 10^3A designation were obtained from Waters. The 100A column was always last in series when a column set was used. Polystyrene standards were obtained from Waters (MW=1.35K, 4K, 17.5K, 50K, 110K, 250K, 390K, and 2700K) and Toyo Soda, USA (Atlanta, GA) (2.8K). Orthodichlorobenzene, styrene monomer, and normal hydrocarbons were also used for calibration. These materials, as well as the various test solutes, were purchased from a variety of suppliers. HPLC grade THF (UV-stabilized), toluene and chloroform were obtained from Waters and degassed before use.

<u>Sample Preparation.</u> Calibration standards and test solutes were injected as dilute solutions in the eluent. Polystyrene standards were 0.03% (w/v). Styrene, ODCB and normal hydrocarbons were 0.15% (w/v), except for dodecane and tridecane (0.65%). Samples involving more complex matrices were prepared by crushing (if necessary), dissolving a weighed amount in the eluent, and filtering through a 0.45 Millex-SR filter cartridge (Millipore, Bedford, MA).

<u>Data Analysis.</u> For the determination of PSDs, the calibration data was converted to R vs log Ø by a BASIC computer program

"PORESIZE", executed on an Apple II Plus microcomputer. Plotting of the differential and cumulative PSDs was done via APPLE PLOT (Apple, Cupertino, CA). Other programs were written in BASIC to calculate Gaussian envelopes given a mean and standard deviation ("GAUSS"), to sum two or more Gaussians in any desired proportions ("COADD"), to integrate the summed distribution ("BOXES") and to transform coordinates from [log $\emptyset$, R] to [V_E, log (MW)].

Results and Discussion

Solvent Effects. The eluent in GPC is deliberately chosen to be a strong solvent for the solute so that retention by mechanisms other than size exclusion (e.g. adsorption) will not occur to an appreciable extent. Therefore, the choice of solvent is not expected to greatly influence the chromatographic results. Solvent effects of two kinds do occur in practice. The first is detector related. In the separation of normal hydrocarbons by GPC with THF as eluent, some peaks are positive and some are negative when detection is by differential refractometry. (See, for example, Figures 1-2 of reference 6). Since the lighter hydrocarbons have a refractive index less than that of THF, peaks for less than C10 are negative. The higher hydrocarbons show a positive response which complicates quantitation of the peaks, especially near the crossover from negative to positive. This sort of solvent effect can generally be avoided by the selection of another eluent. In the case of n-hydrocarbons, the use of toluene affords a chromatogram in which all the peaks are negative.

The second type of solvent-related effect which commonly occurs is observed when a mixture of 1-octanol and 1,8-octanediol is analyzed in two different eluents. In chloroform, the two alcohols are not resolved due to their similar molecular size. In THF, however, resolution nearly to baseline can be achieved due to differentiation of the alcohols on the basis of hydrogen bonding interactions with THF. Octanediol, having two sites for interaction, forms a species with a significantly larger effective size in solution than does octanol which has only one site for interaction. The separation is therefore enhanced.

Due to the popularity of THF as an eluent for GPC, this sort of "differential solvation" must be kept in mind, particularly when polar solutes are analyzed. This effect can also work against resolution in SMGPC, as demonstrated in Figure 1. Here BHA and BHT are fully resolved in CHCl$_3$ but coelute in THF. Both BHA and BHT have phenolic sites, but the site on BHT is sterically hindered and apparently does not form a hydrogen bond with THF. The hydrogen bonded BHA/THF complex which does form is apparently similar in size to BHT with the result that the two solutes coelute in THF. In chloroform, BHA and BHT are resolved due to significant differences in the molecular sizes of these solutes.

Column Efficiency. The peak capacity (13) for a GPC column used in the analysis of small molecules is related to the number of theoretical plates (N) according to:

$$n = 1 + \frac{\sqrt{N}}{4} \, \Delta \ln V_E \qquad (8)$$

where $\Delta \ln V_E$ specifies the elution range of interest. Since the ULTRASTYRAGEL family of columns offers almost a three-fold increase in N compared to the STYRAGEL family, a 50-70% increase is expected in the number of resolvable peaks per chromatogram. This extra resolving power makes it possible to perform a variety of SMGPC separations on single ULTRASTYRAGEL columns. Examples from application areas including foods, pesticides, pharmaceuticals, and polymer additives have recently been reported (10-12). Two advantages of SMGPC over other separation techniques (e.g. reversed-phase HPLC) are frequently observed: simple preparation of complex samples, and good chromatographic resolution of analyte peaks from interfering species.

Two representative examples of single-column SMGPC separations are presented in Figures 2 and 3. The sample for Figure 2 was a rodent bait from which the active ingredient warfarin was to be determined. Quantitation of this component by SMGPC was shown to be as reliable as for the reversed-phase method which is commonly used (35), with the advantage of a several-fold faster sample clean-up (12).

In Figure 3, the active steroid (triamcinolone acetonide) and preservative (benzyl alcohol) are determined from a steroid cream. The higher molecular weight components of the cream base are well separated from the analytes. The ability to elute all the components of a cream or ointment in a SMGPC analysis gives an important sample preparation advantage over competing separation techniques.

Calibration Curve. The calibration curves for GPC columns can provide some guidance in the selection of a column which would give the best resolution for a given analysis. Figure 4 presents calibration curves for typical 100A, 500A, and 10^3A ULTRASTYRAGEL columns, based on the elution behavior of polystyrene standards and n-hydrocarbons. The slope of the linear portion of each curve is related to the resolving power of the column in that a shallower slope will yield a larger ΔV_E for the same ΔMW. There is a tradeoff between the slope of a calibration curve and the range of molecular weights

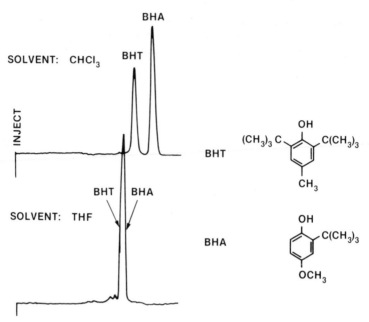

Figure 1.   Separation of BHA and BHT on a 100A Ultrastyragel
column using chloroform and tetrahydrofuran..   Conditions:
1 mL/min; and 254 nm.

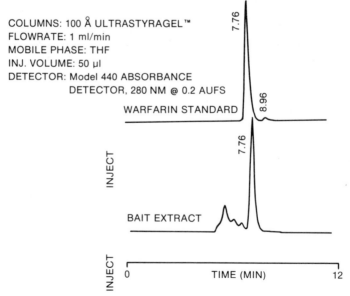

Figure 2.   Determination of warfarin from grain bait by
SMGPC.

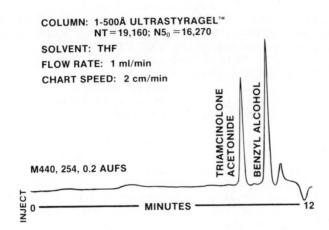

Figure 3. Determination of triamcinolone acetonide and benzyl alcohol from a steroid cream.

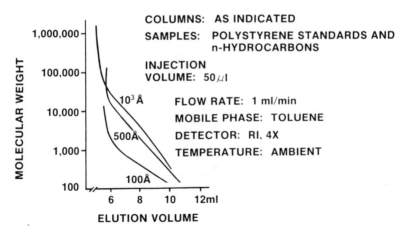

Figure 4. Typical calibration curves for 100A, 500A, and 10^3A Ultrastyragel columns.

resolvable by the column. This is apparent in Figure 4, where the column with the shallowest slope (100A) also covers the narrowest molecular weight range (MW=50 to 1500). For molecules which elute within this relatively narrow range, the 100A column is preferred on the basis of slope. A more extensive range of molecular weights is served by the 500A (MW = 100 - 10,000) and 10^3A (MW = 200 to 30,000) columns. The interplay between column efficiency and the slope of the calibration curve (eq. 1) should not be overlooked. A shallow slope gives a better resolution of peak centers, but for an inefficient column the peaks will be broad and significant overlap (poor resolution) may still occur.

Since the GPC separation is based on effective size in solution of the solutes, and not on molecular weight, conclusions drawn from Figure 4 will apply strictly only to polystyrene standards and n-hydrocarbons. Other compounds may exhibit a different relationship between molecular weight and size. This problem has received attention from several groups (37-41). However, in the absence of a method for assigning a polystyrene-equivalent molecular weight to each solute, the inspection of Figure 4 provides a starting point in the selection of the column which is appropriate for an analysis.

Prediction of Calibration Curves. To address a wider range of molecular weights than is possible with any one column, several columns may be joined in series. If calibration data is available for each individual column, it would be convenient to predict or calculate the calibration curve for the column set. Some rough predictions could be made on the basis of Figure 4. A more accurate answer could be obtained if the individual columns in the set have been calibrated with the same standards. In this case, simply summing the elution volumes of a given standard for each column can give a useful empirical prediction of the elution behavior of the same standard on the column set. Our experience with several examples indicates that this approach can provide accurate results.

A very different scheme for the prediction of calibration curves is presented schematically in Figure 5. This approach invokes a simple theoretical model for the GPC elution process (eq. 5,6). The example to be discussed in this case is a column set composed of two 500A and one 10^3A ULTRASTYRAGEL columns. Calibration data for the individual columns is first transformed according to Equations 2 and 3. The resulting cumulative PSDs are fit to Gaussian distributions as shown in Figure 6 for the 500A column. The area under each Gaussian PSD is made proportional to the experimentally determined pore volume of the column. The Gaussian profiles are then added point-by-point to give the overall PSD of the column set. This curve is

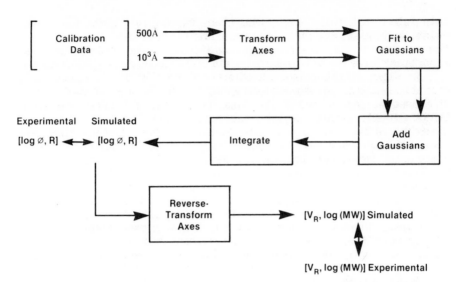

Figure 5. Scheme for prediction (simulation) of the calibration curve for a column set consisting of two 500A and one 10^3A Ultrastyragel columns (see text for details).

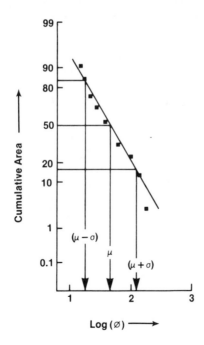

Figure 6. Plot on probability paper of cumulative PSD data for a 500A Ultrastyragel column. The mean (μ = 1.70) and standard deviation (σ = 0.42) of the Gaussian PSD were determined graphically.

integrated to yield the predicted cumulative PSD. At this
point, a comparison with the experimental calibration curve for
the column set can be made, provided that the column set data is
also transformed according to Equations 2 and 3.

In Figure 7, the predicted cumulative PSD is compared with
the actual curve for the column set. It should be noted that
the prediction is based on data from one 500A and one 10^3A
ULTRASTYRAGEL column which had been calibrated in toluene.
(Previous work (19,42) has demonstrated that equivalent PSDs are
obtained with several eluents including chloroform, methylene
chloride, THF, and toluene.) Neither of these columns was
included in the actual column set, which was independently
calibrated in THF using a different instrument. Reasonably
close agreement between prediction and experiment is observed in
Figure 7. The predicted curve is shifted slightly to the right
throughout the range, and would follow the experimental points
very accurately if the shift were eliminated. The cause of the
shift has not been determined, but instrumental differences
(e.g. calibration of flow rate) could provide the explanation.

Alternatively, the predicted cumulative PSD can be converted
to a conventional calibration curve for comparison with
experimental results. This requires reversing the
transformation of equations 2 and 3. To calculate values of
V_E based on values of R, predicted values of V_{EX} and V_{IN}
are needed. The most reasonable approach is to predict each of
these as the sum of the values for the individual columns. The
conversion of the cumulative PSD into a predicted conventional
calibration curve may introduce additional inaccuracy through
the prediction of column set values for V_{EX} and V_{IN}. As
Figure 8 demonstrates, this problem is not severe, and a good
prediction results.

As another example of this approach, the calibration curve
for a column set containing one 100A and one 10^3A
ULTRASTYRAGEL column was predicted with the results shown in
Figure 9 (cumulative PSD) and Figure 10 (calibration curve).
For this example, the columns used for prediction were also used
in the column set, with all measurements made on the same
instrument. Here the good agreement between the predicted and
experimental cumulative PSD (Figure 9) is lost to some extent
upon conversion to the conventional calibration curve (Figure
10). The predicted exclusion volume is in error by nearly a
milliliter, for reasons which are not yet clear. This error can
be removed if the experimentally determined column set value for
V_{EX} is used in generating the predicted calibration curve. As
expected, the result of this substitution is an improved fit to
the experimental data for the high molecular weight region of
the curve.

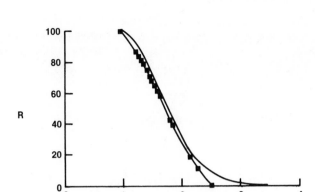

Figure 7. Predicted (smooth curve) and experimental (boxes) cumulative PSDs for a column set consisting of two 500A and one 10³A Ultrastyragel columns.

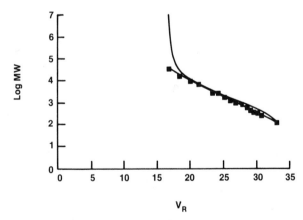

Figure 8. Predicted (smooth curve) and experimental (boxes) calibration curves for a column set consisting of two 500A and one 10³A Ultrastyragel columns.

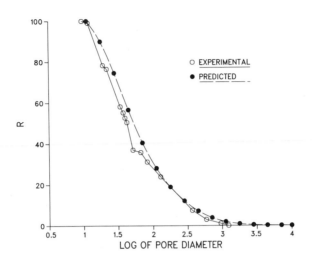

Figure 9. Predicted and experimental cumulative PSDs for a column set consisting of one 100A and one 10^3A Ultrastyragel columns (see legend).

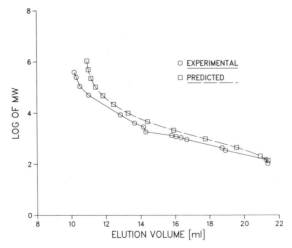

Figure 10. Predicted and experimental calibration curves for a column set consisting of one 100A and one 10^3A Ultrastyragel column (see legend).

Gaussian PSDs. The two examples discussed above indicate the feasibility of the scheme outlined in Figure 5. Additional testing is needed to assure the validity of this approach. Nonetheless, the ability to accurately predict calibration curves on the basis of a simple model implies that the Gaussian PSDs generated by this method are realistic representations of the range of pores available in the columns studied. The PSDs are log normal as of function of $\emptyset$ (eq. 2). Some workers have previously treated PSDs as log normal based on molecular weight (13,43). Figures 11 and 12 demonstrate that this view does not apply to PSDs which are measured by GPC. In Figure 11, calibration data for a 10^3A column was transformed according to Equations 2 and 3 and then fit to a Gaussian. The integrated Gaussian yields a predicted cumulative PSD which compares favorably with experimental data. For Figure 12, only the V_E values were transformed. A Gaussian was fit and integrated as before. The predicted cumulative PSD in this case shows a significant deviation from experiment throughout the molecular weight range, indicating that Equation 2 is necessary to obtain a useful Gaussian PSD.

The success of this prediction scheme suggests several opportunities. It should be possible to specify the desired characteristics of a calibration curve in advance (e.g. linear from MW = 10^2 to 10^5) and then predict the proper combination of available packing materials which would yield those characteristics either in a column set or a single mixed-bed column. Alternatively, the best combination of PSDs could be predicted as an aid in decisions regarding the design of new packing materials. This approach to the prediction of calibration curves may also suggest an interesting alternative to linear and polynomial curve fitting of GPC calibration data. Work is presently underway in our laboratory to pursue these opportunities.

Conclusion

For the optimal application of GPC to the separation of discrete small molecules, three factors should be considered. Solvent effects are minimal, but may contribute selectivity when solvent-solute interactions occur. The resolving power in SMGPC increases as the square root of the column efficiency (plate count). New, efficient GPC columns exist which make the separation of small molecules affordable and practical, as indicated by applications to polymer, pesticide, pharmaceutical, and food samples. Finally, the slope and range of the calibration curve are indicative of the distribution of pores available within a column. Transformation of the calibration curve data for individual columns yields pore size distributions from which useful predictions can be made regarding the characteristics of column sets.

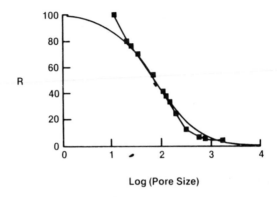

Figure 11. Predicted (smooth curve) and experimental (boxes) cumulative PSD for a 10^3A Ultrastyragel column.

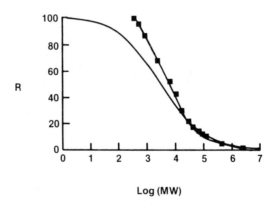

Figure 12. Predicted (smooth curve) cumulative PSD for a 10^3A Ultrastyragel column, determined incorrectly due to failure to convert log (MW) values to log 0. Experimental cumulative PSD (boxes) is shown for comparison.

Acknowledgments

The authors thank Mark Andrews, John Morawski, and Alex Newhart
for performing some of the separations discussed in the text
(Figures 1-2), and gratefully acknowledge the assistance of
Janet Newman in the preparation of this manuscript.

Literature Cited

1. Cortis-Jones, B. Nature 1961, 79, 731.
2. Cazes, J.; Gaskill, D.R. Sep. Sci. 1967, 2, 421.
3. Bombaugh, K.J.; Dark, W.A.; Levangie, R.F. Anal Chem.
    1968, 236, 443.
4. Conroe, K.E. Chromatographia, 1975, 8, 119.
5. Vivilecchia, R.V.; Lightbody, B.G.; Thimot, N.Z.; Quinn,
    H.M. J. Chromatog. Sci. 1977, 15, 424.
6. Krishen, A. J. Chromatog. Sci. 1977, 15, 434.
7. Walter, R.B.; Johnson, J.F. J. Liq. Chromatogr. 1980, 3,
    315.
8. Schultz, H.S.; Ekmanis, J.L.; Tisdale, V.R.; Baptiste, A.J.;
    Crossman, L.W. paper presented at Pittsburgh Conference on
    Analytical Chemistry and Applied Spectroscopy, Atlantic
    City, N.J. March 8-13, 1982, Abstract No. 392.
9. Schultz, H.S.; Alden, P.G.; Ekmanis, J.L. paper presented at
    Pittsburgh Conference on Analytical Chemistry and Applied
    Spectroscopy, Atlantic City, N.J. March 7-12, 1983, Abstract
    No. 393.
10. Richardson, H.; Tarvin, T.L. paper presented at Pittsburgh
    Conference on Analytical Chemistry and Applied Spectroscopy,
    Atlantic City, N.J. March 7-12, 1983, Abstract No. 571.
11. Morawski, J.; Cotter, R.L.; Ivie, K. paper presented at
    Pittsburgh Conference on Analytical Chemistry and Applied
    Spectroscopy, Atlantic City, N.J. March 7-12, 1983, Abstract
    No. 890.
12. Andrews, M.W.; Morawski, J; Newhart, A.T. paper presented at
    Pittsburgh Conference on Analytical Chemistry and Applied
    Spectroscopy, Atlantic City, N.J. March 7-12. 1983, Abstract
    No. 951.
13. Yau, W.W.; Kirkland, J.J.; Bly, D.D. "Modern Size Exclusion
    Liquid Chromatography" Wiley: New York, 1979; Chap.2,4.
14. Majors, R.E. J. Chromatog. Sci. 1977, 15, 334.
15. Yau, W.W.; Kirkland, J.J.; Bly, D.D.; Stoklosa, H.J. J.
    Chromatogr. 1976, 125, 219.
16. Schultz, H.S.; Alden, P.G. paper presented at
    Pittsburgh Conference
    on Analytical Chemistry and Applied Spectroscopy, Atlantic
    City, N.J. March 7-12. 1983, Abstract No. 955.

17. Schultz, H.S.; Alden, P.G.; Ekmanis, J. paper presented at
    185th ACS National Meeting, Seattle, WA, March 20-25, 1983,
    Abstract No. ORPL 200.
18. Halasz, I. Ber. Bunsenges Phys. Chem. 1975, 79, 731.
19. Halasz, I.; Martin, K. Angew. Chem., Int. Ed. Engl. 1978,
    17, 901.
20. Halasz, I.; Vogtel, P. Angew. Chem., Int. Ed. Engl. 1980,
    19, 24.
21. Werner, W.; Halasz, I. Chromatographia 1980, 13, 271.
22. Nikolov, R.; Werner, W.; Halasz,I. J. Chromatog. Sci.
    1980, 18, 207.
23. Werner, W.; Halasz, R. J. Chromatog. Sci. 1980, 18, 277.
24. Groh, R.; Halasz, I.; Anal. Chem. 1981, 53, 1325.
25. Crispin, T.; Halasz, I. J. Chromatogr. 1982, 239, 351.
26. Freeman, D.H.; Poinesca, I.C. Anal. Chem. 1977, 49, 1183.
27. Schram, S.B.; Freeman, D.H. J. Liq. Chromatogr. 1980, 3,
    403.
28. Freeman, D.H.; Schram, S.B. Anal. Chem. 1981, 53, 1235.
29. Kuga, S. J. Chromatogr. 1981, 206, 449.
30. Warren, F.V.; Bidlingmeyer, B.A. submitted to Anal. Chem.
31. Ekmanis, J.L. unpublished data.
32. Cantow, M.J.R.; Porter, R.S.; Johnson, J.F. J. Polym. Sci.,
    Part A-1 1967, 5, 987.
33. Cantow, M.J.R.; Johnson, J.F. J. Polym. Sci., Part A-1
    1967, 5, 2835.
34. DeVries, A.J.; LePage, M.; Beau, R.; Guillemin, C.L. Anal.
    Chem. 1967, 39, 935.
35. AOAC 6.141-2, 13 ed.
36. Smith, W.B.; Kollmansberger, A. J Phys. Chem. 1965, 69,
    4157.
37. Hendrickson, J.G.; Moore, J.C. J. Polym. Sci., Part A-1
    1966, 4, 167.
38. Hendrickson, J.G. Anal. Chem. 1968, 40, 49.
39. Lambert, A. J. Appl. Chem. 1970, 20, 305.
40. Lambert, A. Anal. Chim. Acta. 1971, 53, 63.
41. Krishen, A.; Tucker, R.G. Anal. Chem. 1977, 49, 898.
42. Engelhardt, H., personal communication.
43. Yau, W.W.; Ginnard, C.R.; Kirkland, J.J. J. Chromatogr.
    1978, 149, 465.

RECEIVED September 29, 1983

High-Performance High-Speed Gel Permeation Chromatography

A Systems Approach

RONALD L. MILLER and JACK D. KERBER

The Perkin-Elmer Corporation, Main Avenue, Norwalk, CT 06856

The tremendous advances in size-exclusion column technology in the last decade have resulted in an order of magnitude reduction in analysis times in gel-permeation chromatography (GPC) since the technique was first introduced in the 1960's. The availability of highly efficient (up to 50,000 plates/meter) columns containing a broad pore-size distribution has enabled many separations to be performed using a single column, with no loss of resolution. The more recent development of 5-μm polystyrene-divinylbenzene gel packings has resulted in capabilities for oligomer separations which were unheard of just a few years ago. As GPC separations are performed in less time, with fewer columns, the performance of other components of the chromatographic system becomes critical. A well-designed system for high-resolution, high speed GPC should embody precise control of flow rate and column temperature, minimal peak-broadening effects from both extra-column sources and the columns themselves, and sophisticated data acquisition and processing. The separation of oligomers is an application which clearly demonstrates the advantages of a systems approach to high-resolution, high-speed GPC.

Since the introduction of gel-permeation chromatography (GPC) in the 1960's, there have been tremendous advances in polymer gel size-exclusion column technology. Polystyrene-divinyl benzene copolymer gels, and the techniques by which they are packed into columns, have improved to the point where commercial columns exhibit up to 50,000 plates/meter. These 10-μm gels are sufficiently rugged to permit flow rates of up to 3.0 ml/minute

0097–6156/84/0245–0189$06.00/0
© 1984 American Chemical Society

with low viscosity GPC solvents such as tetrahydrofuran (THF),
with little or no impact on efficiency or column lifetime. The
more recent development of 5-μm gels has resulted in columns with
efficiencies of up to 80,000 plates/meter, usable at flow rates
of up to 2.0 ml/minute with low viscosity solvents. The last
decade has seen an order-of-magnitude increase in efficiency of
GPC columns, which means a three-fold resolution increase for the
same number and length of columns, or alternatively, the ability
to generate equivalent resolution in a fraction of the total
column length. Separations of low-molecular-weight materials may
be performed in minutes using the 5-μm gels, rather than hours,
with a separation power unheard of just a few years ago.

Of course, resolution is not the only criterion for
determining the optimum column set for a given separation: the
column set must cover the molecular-weight range of the sample as
well. A second development which minimized the number of GPC
columns needed for many separations was the development of
columns packed with a mixture of different pore-sized gels.
Operating ranges of these columns span four to five orders of
magnitude in molecular-weight units. The result is that most GPC
separations can be adequately performed with either a single
"mixed-bed" column, or a column set consisting of a mixed-bed
column plus a second column geared to the molecular-weight range
of the sample of interest. The net result is that GPC
separations which required several hours to perform in the 1960's
can now be performed in 15 to 20 minutes in most cases, and in 6
to 10 minutes in some cases, with better resolution than could be
previously achieved. High-resolution, high-speed GPC has thus
acquired a whole new meaning.

The benefits of this advanced column technology cannot be
fully realized without corresponding evolution of other
capabilities of the chromatographic system, however. Because the
time scale of the separation is drastically shortened, factors
such as constancy and reproducibility of temperature and mobile
phase flow rate become much more important. As the contribution
to peak broadening is lessened, extra-column contributions become
more significant. More data must be taken, and taken faster;
manual calculation of molecular-weight averages has already
become obsolete. The increasing availability of and dependence
on the laboratory microcomputer for GPC calculations has spurred
development of powerful software tools using computer
graphics to provide a visual dimension to GPC data reduction.

A systems approach to high-resolution, high-speed GPC takes
all of these factors into consideration. Several aspects are
worthy of detailed discussion.

Modern GPC Columns

The heart of a GPC separation system is, of course, the columns.
As has been previously stated, column efficiencies have greatly
improved over the last decade. The high efficiency of today's
GPC column provides a better separation in less time. Nowhere is
this more apparent than in applications which require the
separation of oligomers. Low-molecular-weight condensation
polymers often fall into this class. The analyst can, via
judicious column selection, gain very high resolution in a
reasonably short time frame. The separation of Figure 1 was
obtained using only two 30-cm columns packed with 10-μm gels,
eluted with tetrahydrofuran (THF) at 1.0 ml/minute. This degree
of separation, using only 60 cm of column length, was not
possible a few years ago, when column efficiencies of 5,000 to
7,000 plates represented the state of the art; more columns or
recycle would have been required.

Figure 2 illustrates a separation carried out using a
single 5-μm gel column, eluted with THF at a flow rate of 1.5
ml/minute. Excellent resolution is obtained with a single
column; the last two compounds to elute differ by only 28
molecular-weight units. The resolution shown in Figure 2
requires a column efficiency of over 20,000 plates, generated
between 4.5 and 6 minutes, or about 80 plates/second.
Efficiencies of 24,000 plates at permeation and 23,000 plates at
total exclusion were measured at flow rates of 1.0 ml/minute for
this column. Separation speed is quite good also, but does not
represent the limit which can be attained.

Bandwidth. Column efficiency may also be expressed in terms of a
bandwidth. The bandwidth is defined as the volume of mobile
phase containing 95% of an eluted compound, or, equivalently,
four standard deviations of a statistical distribution of the
same shape as the chromatographic peak:

$$\text{Bandwidth} = 4\ \sigma\ =\ 4\ V_R\ (N)^{-\frac{1}{2}} \tag{1}$$

Equation 1 shows that bandwidth is merely a means of expressing
column efficiency, N, as a function of elution volume, V_R.
Assuming an exclusion volume of 5 ml per column allows
construction of Table I from Equation 1. Table I lists the
bandwidth in microliters as a function of column plate number and
the number of columns in series. The data assume that the plate
number may be generated at total exclusion, as well as at total
permeation; actual measurements made using the smaller pore size
column substantiate this.

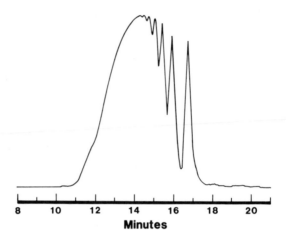

Figure 1. Separation of epoxy cresol Novolac oligomers.
Columns: Perkin-Elmer/PL gel 10-μm 100 A and 10-μm 1000 A.

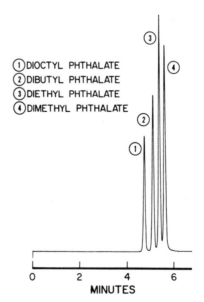

Figure 2. GPC separation of phthalate esters. Column:
Perkin-Elmer/PL gel 5-μm 100 A.

Table I. Bandwidths (µl) of GPC columns at total exclusion

Plates per Column	Number of Columns			
	1	2	3	4
12,000	183	258	316	365
16,000	158	224	274	316
20,000	141	200	245	283
24,000	129	183	224	258

Polymer gel GPC columns packed with 10-µm gels can exhibit efficiencies of 12,000 to 16,000 plates depending on the pore size. Single columns of this type produce bandwidths from 160 to 180 µl. As columns are coupled in series, bandwidth increases as the square root of the number of columns, as may be seen from Equation 1. Plate number doubles, but so does the exclusion volume. The 5-µm gel columns typically achieve 20,000 to 24,000 plates, and are represented by the bottom two rows of the table. The implications of the bandwidth values in Table I will be discussed below.

Separation Speed. Figure 3 shows chromatograms of polystyrene standards eluted with THF from a GPC column packed with a mixture of different porosity particles, the so-called "mixed-bed" column. A single column of this type covers a sufficiently broad molecular-weight range so that it alone may be used for many analyses. Furthermore, since the resistance to flow for a single column is low, higher mobile-phase flow rates may be used without generating an excessive pressure drop across the column. Figure 3 shows the separation of standards at flow rates of 1.0 and 3.0 ml/minute. The column generated 13,000 plates at the higher flow rate, compared with 12,900 at 1.0 ml/minute. Number-average and weight-average molecular weights of a polydisperse polystyrene sample run at flow rates of 1.0, 2.0, and 3.0 ml/minute were observed to vary by less than 2% when computed against calibrations obtained at the same flow rate. The variation of the molecular-weight averages with flow rate appears to be well within reason, particularly when no attempt was made to thermostat the column during these experiments; the differences between the molecular weights could easily be a consequence of small changes in column temperature between calibration and running the samples.

At high flow rates, the diffusion rates of macromolecules limit the resolution obtainable. This is apparent from Figure 3; the resolution between early eluting peaks suffers as the flow rate is increased. The resolution near the permeation limit is not greatly affected, however, and the effect on calculated molecular-weight averages was observed to be small even for large molecules. What is significant is that separation speed is limited by the nature of the sample, and not by the column.

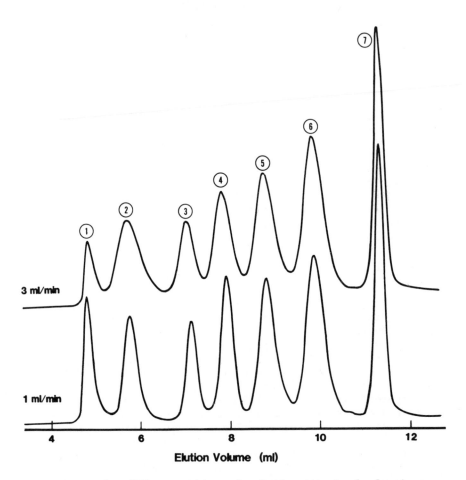

Figure 3. GPC separation of polystyrene standards at
different flow rates. Column: Perkin-Elmer/PL gel 10-μm
mixed.

Instrumental Band Broadening

The data in Table I illustrates that very high-efficiency GPC
columns separate compounds with a minimum of dilution. This is
just another way of expressing column efficiency; the greater the
plate number, the lower the dilution at a given retention volume.
When the peak dilution from the column is small, i.e. when a
small number of highly efficient columns are used, the degree to
which other system components contribute to peak dilution and
broadening becomes much more significant. DiCesare et.al. ($\underline{1}$)
have discussed extra-column contributions to bandwidth for very
high speed reversed-phase liquid chromatography; most of the same
considerations apply to GPC as well.

   In chromatographic systems, the various contributions to
peak broadening are generally independent. This means that the
variance of the system is the sum of the variances from each
contribution. Combining this relationship with Equation 1 yields
an expression for the system bandwidth:

$$\text{Bandwidth}^2 = (4\ \sigma_{total})^2 = \sum_i (4\ \sigma_i)^2 \qquad (2)$$

The peak broadening for the entire chromatographic system,
columns plus the instrument, may thus be estimated from the
bandwidth contribution of each component of the system. The
effective plate number of the system may then be calculated from
Equation 1.

Effect of Injection Volume. Table II shows the effect of
injection volume on peak broadening and measured column
efficiency. The bandwidths listed in Table II are due to
injection volume alone, and were measured using an injector
connected directly into the flowcell of a low-bandwidth detector.
The plate reductions were then calculated for a 24,000 plate
column, such as that represented by the bottom line of Table ·I,
assuming 5 and 10 ml, respectively, for exclusion and total
permeation volumes. Efficiencies of 23,000 plates at exclusion
and 25,000 plates at permeation were actually measured for the
column indicated in Table II. The effect of large injection
volumes is thus to lose 25 to 50% of the potential column
efficiency.

   The injection volume chosen for analysis must represent a
compromise between the amount of sample needed to properly detect
the eluting material, and the amount of extra-column dispersion
the analyst is willing to tolerate. It is also important that
the same injection volume be used for both samples and standards,
and that sample injection be properly synchronized with the start
of data acquisition.

Table II. Effect of injection volume on bandwidth and realized efficiency of a Perkin-Elmer/PL Gel 5- m 100 Angstrom column eluted with THF at 1.0 ml/min

Injection Volume, μl	Injector Bandwidth, μl	Plate Reduction at Exclusion	Plate Reduction at Permeation
3.0	30	1,132	320
6.0	31	1,204	341
10.0	36	1,595	458
23.5	47	2,592	770
50.0	86	6,863	2,397
100.0	160	13,696	6,659

Effect of Tubing Diameter. The contribution of the connecting tubing in the system to the bandwidth can also be estimated. A typical chromatograph might employ about 80 cm of connecting tubing between the injector and the detector. The bandwidth of 80 cm of .007-inch i.d. tubing has been determined to be about 31 μl, (1) equivalent to that due to a 6-μl injection. It may be shown that the bandwidth of the connecting tubing is porportional to the square root of the length and at least the square of the inside diameter. (2) The bandwidth due to 80 cm of .015-inch i.d. tubing is more than 140 μl, a contribution nearly as large as that for a 100-μl injection.

Effect of Detector Flowcell. The same considerations may be applied to the detector flowcell. For example, DiCesare et. al. (1) determined that an 8-μl flowcell in a "conventional" UV detector might have a bandwidth of 70 μl or more. The UV detector employed in Figures 2 and 3 (LC-85B, Perkin-Elmer) has a 1.4-μl flowcell with a bandwidth below 5 μl. In a GPC system employing a single 24,000 plate column, a detector with a 70-μl bandwidth would degrade efficiency by about 24% at exclusion, while the effect of the 1.4-μl flowcell of the detector used in this work is negligible. Refractive index detectors typically have higher bandwidths, ranging from 25 to 100 μl or more for commerical instruments.

Figure 4 illustrates the advantages of optimizing the GPC system with respect to instrumental band broadening. The lower chromatograms were obtained from a "conventional" chromatographic system employing a 10-μl loop injector, about 80 cm of .015-inch i.d. connecting tubing, and a UV detector with an 8-μl flowcell (LC-75, Perkin-Elmer). The sample is a liquid polystyrene resin separated using first a 10-μm gel column (Perkin-Elmer/PL Gel 10-μ m 100 A) and then a 5-μm gel column (Perkin-Elmer/PL Gel 5-μm 100 A) of the same porosity. The same sample was separated on an optimized system, which employed a 6-μl loop injector, 80 cm of .007-inch i.d. tubing, and a UV detector with a 1.4-μl flowcell (LC-85B, Perkin-Elmer), producing the two upper

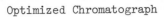

Optimized Chromatograph

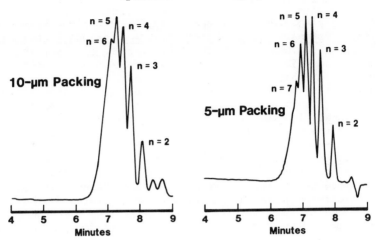

Conventional Chromatograph

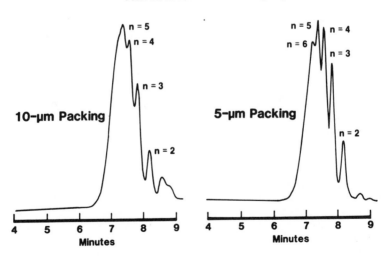

Figure 4. Separation of liquid polystyrene resin on different chromatographic systems. System configuration and column type are defined in the text.

chromatograms in Figure 3. The columns used were those used for
the lower chromatograms. All four separations were performed
with a THF mobile phase at 1.0 ml/minute.

The total extra-column bandwidths, calculated from Equation
2, were 195 µl and 44 µl, respectively, for the conventional and
optimized systems. Column efficiencies were 16,000 plates for
the 10-µm gel column and 24,000 plates for the 5-µm gel column;
the column contributions to bandwidth are given in Table I. The
difference in resolution obtainable between the two
chromatographic systems is readily apparent from Figure 4. The
optimized system produces much narrower peaks, and more of them
as additional oligomer are resolved. In terms of required
bandwidth, the extra-column bandwidth of the optimized system is
about a third of that inherent in the 5-µm gel column at total
exclusion, while the conventional system has a bandwidth greater
than that of either column. The high instrumental bandwidth of
the conventional system is largely due to the contribution from
the connecting tubing; the bandwidth of this system could have
been reduced to about 94 µl by substitutung .007 i.d. tubing. A
system bandwidth of 94 µl is still unacceptable for work employing
a single 5-µm gel column, but may be tolerable for separations
employing multiple 10-µm gel columns.

Table III summarizes the results represented by Figure 4.
The bandwidth values in the table are those calculated for the
total system: the instrument plus the column. The values for
number of plates are for the number of plates realized in the
total system. It can be seen that the optimized system does not
greatly impact column efficiency, the total loss in plates being
only about ten percent at total exclusion for a 24,000 plate
column. This is consistent with an instrumental bandwidth equal
to a third of the bandwidth of the column. The conventional
system, with a bandwidth equal to or greater than that of the
column, exhibited a severe loss in realized efficiency,
particularly at or near exclusion.

Table III. Effect of instrumental bandwidth on column efficiency

Inherent Column Efficiency	Conventional System		Optimized System	
	Total System Bandwidth, µl	Effective Plates	Total System Bandwidth, µl	Effective Plates
16,000 plates				
at permeation	372	11,600	319	15,700
at permeation	256	6,300	164	14,800
20,000 plates				
at permeation	344	13,500	286	19,500
at permeation	255	6,900	148	18,200
24,000 plates				
at permeation	324	15,200	261	23,300
at permeation	234	7,300	136	21,500

The data of Table III represent calculated bandwidths and efficiencies. Actual realized efficiencies were measured for the four chromatograms of Figure 4. For the 10-μm gel column, the conventional system produced an effective efficiency of 11,000 plates, compared with an effective efficiency of 16,000 plates for the optimized systems. These values are in excellent agreement with the calculated values shown on the top line of Table III. Similar measurements on chromatograms obtained from the 5-μm gel columns yielded values of 16,000 and 20,000 plates, respectively, for the conventional and optimized systems. This also represents good agreement with calculated effective efficiencies at total exclusion for a 24,000 plate column.

The 5-μm gel GPC columns are seen to produce tremendous efficiencies, but these efficiencies are only realized when the chromatographic system is optimized with respect to bandwidth. This also holds true to a lesser degree for a well-packed 10-μm gel column.

Effect of Detector Response Time. The speed of response of the detector electronics can also affect resolution. Response times can also be expressed as bandwidths by multiplying by the flow rate in the appropriate units. In the previous discussion, this effect was ignored, as the time constant bandwidths were negligible: less than 12.5 μl for either detector. Figure 5 shows an example of what can happen when the time constant bandwidth is too large. The chromatographic system used for the separations shown in Figure 5 is an optimized system incorporating a refractive index detector; bandwidth contributions from the flowcell, tubing, and injector combine to produce a volume bandwidth of 52 μl for this system. The time constants of 5 and 0.5 seconds equate to bandwidths of 167 and 16.7 μl, respectfully, for total system bandwidths of 174 and 55 μl. The effect on resolution is readily apparent; the faster response time produces a total bandwidth acceptable for all applications except where maximum resolution for a single 5-μm gel column is required near total exclusion. The 5 second response time, on the other hand, is of little use except when several of the 10-μm gel columns are used. It is interesting to point out, however, that a system bandwidth of 174 μl was thought to be quite suitable a few years ago, when column efficiencies were significantly lower, and more columns were used.

Data Acquisition and Processing

The data acquisition rate can also contribute to the integrity of GPC data. Chromatograms traced on a recorder are in response to an analog signal, and are continuous traces. Calculation of molecular weights, however, requires digitized data. The frequency of measurement used when digitizing an analog signal is

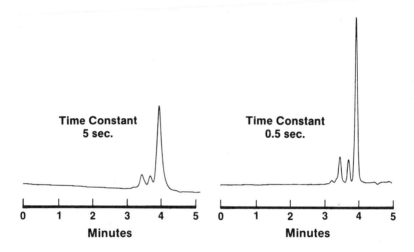

Figure 5. ' The effect of detector time constant on the GPC
separation of a liquid epoxy resin.   Column: Perkin-Elmer/
PL gel 5-μm 100 Angstrom.   Eluent: THF at 2.0 ml/min.
Injection volume: 6 μl.   Detector: LC-25 RI detector
(Perkin-Elmer).

known as the information bandwidth, and can also be converted to volume units. Table IV shows the quantitative effect of the size of the information bandwidth. The sample and conditions are those of Figure 5, except that flow rate was 1.5 ml/minute, and the 0.5 second detector response time was used throughout. Separations of the liquid epoxy resin were performed at various data acquisition rates, and molecular-weight averages calculated against calibration data obtained at a data rate of 0.1 seconds/point.

Table IV. Effect of sampling rate on GPC results

Data Rate, Seconds/point	Time of Largest Data Point, Min.	Molecular-Weight Averages*		
		No. Ave.	Wt. Ave.	Z Ave
0.1	4.93	377	400	435
0.2	4.94	375	398	433
0.5	4.95	372	395	430
1.0	4.97	366	387	420
2.0	5.00	360	379	408
5.0	5.17	328	343	375

* Relative to calibration data taken at 0.1 seconds/point.

The data rates in Table IV correspond to information bandwidths varying from 2.5 µl to 125 µl. The retention times of the largest peak, when taken as the time of the largest data point in the digitized data, show a definite trend, increasing as the time between measurements increases. This is entirely a consequence of the information bandwidth; the analog chromatograms were identical. Table IV also shows the effect on the calculated molecular-weight averages, when calculations were performed relative to calibration data taken from a digitized chromatogram for which a very fast data rate was used. The decreasing molecular-weight averages also varies as the slope of the calibration curve, and would be much greater for a broader range column.

Thus, data rates of 0.5 seconds/point are required to suppress band-broadening contributions from data acquisition. This does not define the limiting requirement of the data system, however. Between 50 and 100 data points are desired to accurately define a molecular-weight average for a single peak, particularly an average representing a higher statistical moment such as the Z-average. The chromatograms of Figure 3 contain seven peaks; 400 to 800 data points are optimum for this chromatogram, when sufficient data points are included to adequately define baseline. The chromatogram of Figure 3 obtained at a flow rate of 3 ml/minute thus requires data rates of 100 to 200 points per minute (300 points per minute were actually used).

Needless to say, processing this data in a time frame
compatible with the time needed for chromatography cannot be done
without the aid of a computer; this is true for even the
20-minute separations of Figure 1. Since the polymer chemist
working with GPC is becoming more dependent on the computer
programmer, it is pertinent to include software as an integral
part of the GPC data system.

The sheer volume of data generated by high resolution, high
speed GPC mandates some type of media storage for raw data,
together with the ability to recall, replot, and rework any of
this raw data. Automation of both data acquisition and data
processing is required to keep pace with the speed at which
samples can be run. This alone may be sufficient for the quality
control laboratory, but the research laboratory also requires the
ability to deal with a particular chromatogram in greater detail
in a more leisurely manner.

The advantages of interactive computer graphics in GPC come
into play here. If the molecular weight averages of two samples
differ, the replotting of chromatograms or distributions on the
CRT of a computer terminal permits a fast, easy comparison of
just how the samples differ. Given appropriate software, screen
graphics can be used to not only redisplay, but also to rescale,
expand, and even subtract chromatograms and distributions, to
plot and manipulate GPC calibrations, and to define baseline and
summation limits to be used in numerical computations. These
capabilities provide a visual dimension not available from mere
numbers, and enable the chemist to solve problems faster and
easier.

The most important consideration of software, however, is
that is must provide the correct answers. This requires that the
appropriate molecular weight be associated with each data point,
which relates to the techniques and algorithms used in
constructing the calibration curve. Calibration curves are
generated from chromatographic data obtained on standards of
known molecular weight; both monodisperse and polydisperse
standards have been used. A discussion of the relative merits of
each technique is beyond the scope of this paper; suffice it to
say that the model used by the software should reflect the true
calibration as closely as possible.

Temperature Control

The importance of temperature control of the GPC column cannot be
overstated. The use of temperatures above ambient results in
lower mobile-phase viscosity, which in turn reduces the back
pressure generated by the column. Column life is prolonged, and
in some cases higher flow rates may be employed. The reduction
in mobile-phase viscosity improves both the rate and efficiency
of mass transfer processes, enhancing column performance. While

the benefits of elevated temperature are certainly desirable, the use of constant temperature is critical. Mark-Houwink coefficients, the parameters which describe the relationship between molecular weight and hydrodynamic volume (and therefore elution volume), are significantly termperature dependent. Polymer solubility improves with increasing temperature; polymer molecules in solution uncoil to a greater degree, and hence occupy larger volume and elute earlier from the GPC column.

The effect of changing temperature on GPC results is illustrated in Table V for a polystyrene sample; a column was calibrated using monodisperse polystyrene standards. The standards and the sample were both run at 25 and 30 C. The detrimental effect of a change in temperature between calibration and sample analysis is obvious; a five-degree C change in temperature was seen to produce errors of 11-14% in the weight-average molecular weight. In this case the sample and the standards were identical chemically. If they are not, their Mark-Houwink coefficients may show differences in temperature dependency, and the errors become compounded. If the goal of the chemist is to attain reproducibility to within 1%, the column temperature must be maintained to within 0.5 C or better throughout the course of the experiments.

Table V. Effect of temperature on weight average molecular weight

Analysis Temperature	Calibration Temperature	Weight-Average Molecular Weight
25 C	25 C	357,000
25 C	30 C	313,000
30 C	25 C	400,000
30 C	30 C	352,000

An air-bath oven is an excellent choice for GPC in that a substantial number of columns may be accomodated by a single unit. Costs are low and temperature stability and reproducibility quite good. Some type of heat-exchange device should be placed in the oven to raise the temperature of the mobile phase to the desired point before it reaches the column; this practice helps eliminate temperature gradients along the column axis. Injectors can be mounted directly on (or even in) an oven, minimizing the amount of heat exchange between the mobile phase and the injector.

Solvent Delivery

Effect of Flow Rate Errors. The effect of flow rate errors on molecular-weight averages calculated from GPC data has been

discussed by Bly, et al.(3)   These workers concluded that flow
rate repeatability of better than 0.3%, flow rate drift of less
than 1% over the time of the chromatogram, and short-term random
variation (noise) of better than 4%, are all required to
reproduce molecular-weight averages to within 6%. Thus, the most
important criteria for a GPC pumping system are, respectively,
resettability, drift, and pulsation. Absolute accuracy of flow
rate must also be considered if comparison of results obtained on
different instruments is also important.

     The exact magnitude of flow-rate induced errors in the
molecular-weight averages depends on the slope of the calibration
curve: the steeper the slope, the more a given flow rate
variation affects reproducibility of the averages. GPC
separations employing a single 25- to 30-cm mixed bed column
probably represent the worst case. Table VI illustrates the
effect of a one percent error in the flow rate resettability for
a column of this type. Using a given set of calibration data and
a given set of raw slice areas for a polystyrene sample,
reference values of the various molecular-weight averages were
computed. The mobile phase was THF at 1.0 ml/minute flow. A
flow rate increase of one percent between the time of calibration
and sample analysis was then simulated by multiplying each of the
retention times in the calibration data set by 1.01, and
repeating the molecular-weight calculations. A decrease in flow
rate was simulated in a like manner. The results indicated that
even a small error in flow rate generates very large errors in
molecular weight, particularly for a column with a steep
calibration curve.

Table VI. Effect of flow rate errors on molecular-weight averages

Molecular Weight	No Change	1% Increase	1% Decrease
Number-Average	125,000	141,700 (+13%)	110,100 (-12%)
Weight-Average	384,900	452,800 (+18%)	331,000 (-14%)
Z-Average	1,621,000	2,628,000 (+62%)	1,078,000 (-27%)

     While the effects of flow rate drift or noise at the one
percent level over the duration of the separation are not nearly
as disastrous as the case illustrated in Table VI, the data serve
to demonstrate the need for flow rate stability and
repeatability. The absolute accuracy of flow rate is of lesser
importance, as this type of variation only manifests itself when
comparing raw data obtained on different instruments, all of
which should be calibrated independently of each other in any
case. It should be pointed out that the GPC calibration should
always be redetermined whenever any component of the system is
changed; this is simply good laboratory practice.

Pumps for GPC. The most important considerations when selecting
a solvent delivery system are those involving flow rate
resettability, drift, and noise. Reciprocating-piston pumps,
either in single-piston or multiple-piston configurations, are by
far the most commonly used solvent delivery devices for GPC. In
a dual-head pump, for example, each pump head operates
essentially 180 degrees out of phase with the other, so that one
pump head is always delivering solvent; pulsation occurs only at
the point of "crossover" between one pump head and the other.
Advanced designs of single-piston pumps minimize pulsation by
refilling the piston at a much faster rate than it delivers
solvent. In either case, some additional pulse-dampening
capability is generally provided to further reduce short term
flow rate fluctuations, or flow rate "noise". Since the pumps
used for GPC tend by and large to be those designed for
"conventional" chromatography, their pulse dampeners may be
optimized for applications producing higher back pressures than
GPC. In this case, placing a flow restrictor between the pump
and injector may improve flow rate reproducibility. In this
work, 3 to 9 meters of coiled tubing, 0.007" i.d., was used as a
flow restrictor. This coil was placed inside the oven, and also
served to preheat the mobile phase.
    A final aspect of GPC solvent delivery relates to the
solvent reservoirs themselves. The ability to perform in situ
helium degassing of solvents, provide inert gas blankets over
solvents, and protect solvents from contamination from external
sources are worth consideration from the standpoints of
convenience and safety alone. If these features are provided
for, it is a small step to also provide a small positive
pressure, say 10 psi or so, to the solvent reservoir. This
positive pressure helps minimize the formation of solvent vapors
in the pump chamber during the refill part of the pump stroke,
and improves the flow rate reproducibility of rapid-refill type
pumps delivering high-vapor-pressure solvents.

System Reproducibility. Table VII describes the reproducibility
achievable with an optimized GPC system. Twelve consecutive
analyses of the same polystyrene sample were analyzed to produce
these data. The pump used was a single-piston rapid-refill type
reciprocating pump (Series 10, Perkin-Elmer) equipped with
reservoir pressurization and restrictor coil as discussed above.
The mobile phase was THF at 1.0 ml/minute, and the reservoir
pressure 11 psi. The column temperature was controlled at 40 C
by placing the column (Perkin-Elmer PL Gel 10-μ MIXED) and the
restrictor coil in an air bath oven (LC-100, Perkin-Elmer) to
reduce any variability due to temperature. Samples were injected
with an autosampler (Model 420B, Perkin-Elmer) containing a
fixed-volume loop injection valve. A variable wavelength UV
detector (LC-75) operating at 265 nm was used as the detector.
Molecular-weight averages were calculated for all twelve

injections using the same baseline times, calibration curve, and
summation limits. The results, summarized in Table VII,
illustrate the precision which can be routinely obtained when all
sources of variation are controlled. Relative standard
deviations lower by about a factor of three have been obtained
using this system for low-molecular-weight polyethoxylated
phenol, separated using a column with a less "steep" calibration
curve.

Table VII. Summary of results from twelve repetative analyses of
polystyrene.

Parameter	Mean	Standard Deviation	Relative Standard Deviation, %
Number-Average Mol. Wt.	140,010	970	0.69
Weight-Average Mol. Wt.	381,500	2,799	0.73
Z-Average Mol. Wt.	1,211,000	13,872	1.15

Summary

We have demonstrated the benefits which can be obtained from
high-efficiency GPC column technology when the chromatographic
system is properly optimized. Band broadening from extra-column
sources must be minimized to realize the full efficiency of
modern GPC columns. Proper control of both flow rate and column
temperature is vital to maximizing reproducibility in GPC.

Literature Cited

1. DiCesare, J. L.; Dong, M. W.; Atwood, J. G. J. Chromatogr.
    1981, 217, 369-85.
2. Martin, M.; Eon, C.; Guiochon, G. J. Chromatogr. 1975, 108,
    229-41.
3. Bly, D. D.; Stoklosa, H. J.; Kirkland, J. J.; Yau, W. W.
    Anal. Chem. 1975, 47, 1810-3.

RECEIVED September 12, 1983

Deuterium Oxide Used to Characterize Columns for Aqueous Size Exclusion Chromatography

HOWARD G. BARTH

Research Center, Hercules Incorporated, Wilmington, DE 19899

FRED E. REGNIER

Department of Biochemistry, Purdue University, West Lafayette, IN 47907

In order to characterize size-exclusion chromatographic (SEC) columns, both the interstitial
volume and the pore volume of a packed column must
be determined. This information is required for
the construction of a calibration curve as well as
to obtain SEC distribution coefficients. In
aqueous SEC, either glucose or deuterium oxide
(D_2O) are commonly used to measure the total
permeation volume of a column. Using LiChrospher
silica packings with a glycerylpropyl silane bonded
phase (SynChropak GPC), we found that the elution
volume of D_2O was significantly greater than the
results obtained for glucose. Controlled-pore
glass packings which have narrower pore-size
distributions did not exhibit this property. From
these results, it appears that the silica packing
contains a population of micropores which are
accessible only to low molecular weight probes.

Size exclusion chromatography (SEC) is a separation process by
which molecules are fractionated by size on the basis of differential penetration into porous particulate matrices. Elution
volume (V_e) of any given molecular species relative to another
of different size is dependent on the pore diameter of the
matrix, pore-size distribution, pore volume (V_i), interstitial
volume (V_o) and column dimensions. Use of SEC to estimate
molecular size is achieved by plotting the log of the molecular
weight of a series of calibrants against their elution volume.
Since V_e is a function of V_o and V_i, its magnitude will be
dependent on the geometry of a column.

A more useful and fundamental parameter than elution volume
is the dimensionless size exclusion distribution coefficient
(K_D) which is related to V_i and V_o by the equation:

0097–6156/84/0245–0207$06.00/0
© 1984 American Chemical Society

$$K_D = \frac{V_e - V_o}{V_i} \qquad (1)$$

Use of K_D instead of V_e in the calibration of columns produces a calibration curve that is independent of column dimensions and pore volume. To obtain K_D for any species requires the determination of V_o and V_i in addition to V_e. V_o is usually taken as the elution volume of an excluded polymer while V_i is equal to $V_T - V_o$. The volume V_T is the total permeation volume of the column and is measured with a low molecular weight compound that totally permeates particle matrices.

Deuterium oxide (D_2O) has been used to determine V_T in SEC columns because its low molecular weight assures high matrix permeation and its high diffusion coefficient is useful in determining column efficiency (1-3). (It should be noted that in aqueous mobile phases, DHO would be present after injecting D_2O into a column because of hydrogen exchange.) In addition to D_2O, tritiated water (THO) has been used as a low molecular weight probe of V_T in SEC (1,4-7). Marsden (4,8), however, cautions that tritium exchange within the crosslinked polysaccharide matrix could result in errors when THO is used to determine V_T. From V_T measurements with $H_2^{18}O$, Marsden found that K_D for THO was 1.09 (8).

The assumption has generally been made in SEC with matrices greater than 100Å pore diameter that there is little, if any, size discrimination of molecules less than 500 daltons, i.e., they would all elute at V_T.

During our studies with SynChropak, a high-performance SEC packing consisting of LiChrospher silica with a glycerylpropyl silane bonded phase, we found to our surprise that the elution volume of D_2O was significantly greater than that of glucose which we had previously used as a low molecular weight calibrant (9-11).

The problem of determining V_T in SEC is similar to that of determining zero retention time (t_o) in other liquid chromatography columns. Recently, there have been several papers dealing with the determination of retention time of a retained peak in HPLC (12-19). In high-performance reversed-phase chromatography, McCormick and Karger (15) and Berendsen, et al., (16) have employed D_2O to measure t_o. Neidhart et al., (12,14) took a different approach by determining the retention times of a solute as a function of temperature. Since the enthalpy of adsorption of a solute onto a stationary phase is negative, the elution time of a retained species should decrease with increasing temperature.

However, none of these methods rigorously examines the possibility that microporosity may also cause differences in t_o between solutes. This paper describes the extent of retention time differences between D_2O and glucose on bonded phase inorganic supports.

Experimental

Apparatus. Pumping systems used in these studies for high-performance columns were a Varian 8500 syringe pump and a Varian 5000 isocratic pump. An Altex 110A was employed for the controlled-pore glass (CPG) columns. Waters Associates model 401 refractometers were used on all instruments. Stagnant mobile phase was kept in the reference side of the refractometer. Samples were injected with a Rheodyne 70-10 injection valve using a 20μl loop (100μl for CPG columns).

Columns. The packing materials were 10μm SynChropak and 37-74μm controlled-pore glass with glyceryl silane bonded phase. SynChropak columns were purchased prepacked in 25 cm x 4.1 mm ID stainless steel columns from SynChrom (Linden, IN). Nominal pore sizes were 100, 300, 1000 and 4000Å.

CPG was dry packed into stainless steel columns using the tap-fill procedure (20). Column dimensions were 100 cm x 4.6 mm ID for the 1000, 1400, 2000 and 3000Å material and 50 cm x 4.6 mm ID for the 75Å packing. A description of these packings is given in Table I. Values listed in the table were obtained from the manufacturer (Electronucleonics Inc.).

TABLE I. GLYCERYL-CPG COLUMN PACKING MATERIAL (200/400 mesh)

Nominal Pore Size, Å	Mean Pore Diameter, Å	Pore Size Distribution, +%	Pore Volume, cc/g	Surface Area, m^2/g
75	75	6.0	0.47	140
1000	1038	7.3	1.22	28
1400	1489	6.4	1.16	17.6
2000	1902	10	0.80	10
3000	3125	10	1.25	7.9

Chemicals. Urea (99+%), glucose and D_2O (99.8%) were obtained from Aldrich Chemical Co. (Gold Label).

Mobile Phase Preparation. Distilled water and 6M urea were filtered under vacuum using a 0.22μm membrane filter (Type GS, Millipore).

Sample Preparation in 6M Urea. Solutions of glucose were prepared directly in 6M urea. D_2O solutions were prepared by diluting equal volumes of D_2O and 12M urea and the resulting solution was then diluted 1:1 with 6M urea.

Elevated Temperature Studies. The Varian 5000 liquid
chromatograph and a Waters Associates 401 differential
refractometer were employed. The column was heated with a
Varian universal heater block at an estimated accuracy of
$\pm$ 0.5°C. About 15-30 minutes were allowed for column
equilibration for a given temperature. The recorder employed
was a Varian 9176.

A 25 cm x 4.6 mm ID long 300Å SynChropak column was used
to evaluate temperature effects. Injections were made with 5%
D_2O and 1.3 mg/ml glucose solutions. D_2O gave a negative
refractive index response.

Because of some peak tailing, the number of theoretical
plates was based on peak width at one-half peak height: N=5.54
$(t_r/w_{1/2})^2$. The pooled standard deviation (all temperatures)
of retention time measurements (df=34) was $\pm$ 0.007 minutes.

Physical Measurements on Supports. Pore diameter and volume
were determined by mercury porosimetry. Micropores were
estimated by the BET and t-curve methods (21, 22).

Results and Discussion

**Elution Volume of D_2O and Glucose on Controlled-Pore Glass and
SynChropak Columns.** The elution volumes of D_2O and glucose on
100, 300 and 4000Å pore-size SynChropak columns are given in
Table II. As indicated, the elution volume of D_2O was greater
than that of glucose in all cases. Because of the smaller
hydrodynamic volume of D_2O, as compared to glucose, this trend
was expected.

However, the sizable elution volume difference between D_2O
and glucose exhibited by the 100 and 300Å columns is
surprising. On the basis of total pore volume, V_i, the
percentage of micropore volume that was available to D_2O and
not glucose was high: 17.4 $\pm$ 1.7% and 8.4 $\pm$ 1.5%, respectively,
for the 100 and 300Å packings. The result obtained with the
4000Å column was within experimental error.

Glucose and D_2O were also tested on five glycerylpropyl
CPG packings of 75, 1000, 1400, 2000 and 3000Å and the results
are presented in Table III. The percentage of micropore volume
that was available to D_2O and not glucose was close to or
within the experimental error of V_e determination for all
columns.

TABLE II. ELUTION CHARACTERISTICS OF D_2O
AND GLUCOSE ON SYNCHROPAK COLUMNS*

Pore Diameter	100Å	300Å	4000Å
D_2O, V_r (ml)	2.58	2.82	2.62
Glucose, V_r (ml)	2.34	2.68	2.60
Δ, ml	+0.24	+0.14	+0.02
V_i, ml**	1.38	1.66	1.47
Micropore volume, %***	17.4±1.7	8.4±1.5	1.4±1.7

* Chromatographic conditions: Mobile phase: H_2O; Flow: 0.5
   ml/min; Chart Speed: 1 in/min; Volume injected: 20μl;
   Sample concentrations: 1 mg/ml glucose and 5% D_2O;
   Columns: 25 cm x 4.1 mm ID; RI detector sensitivity: X4.
** V_i = V_T - V_o where V_T is the elution volume of
   D_2O. For 4000Å columns, V_o = 0.35 ($\pi \cdot r^2 \cdot L$).
   For 100 and 300Å columns, V_o was obtained from 2 x 10^6
   dalton dextran (1.20 and 1.16 ml, respectively).
*** Propagated error assuming flow rate precision of ± 1%.

TABLE III. ELUTION CHARACTERISTICS OF D_2O
AND GLUCOSE ON GYCERYL – CPG COLUMNS*

Pore Diameter	75Å	1000Å	1400Å	2000Å	3000Å
D_2O, V_r(ml)	5.75	14.25	14.20	13.70	13.38
Glucose, V_r (ml)	5.65	14.15	14.18	13.65	13.30
Δ, ml	0.10	0.10	0.02	0.05	0.08
V_i, ml**	2.15	8.43	8.38	7.88	7.56
Micropore volume, %***	4.6±2.7	1.2±1.7	0.2±1.7	0.6±1.8	1±1.8

* Chromatographic conditions: Mobile phase: 0.5 M NaOAc;
   Flow: 0.5 ml/min; Chart Speed: 0.5 cm/min; Volume injected:
   100μl; Sample concentrations: 2 mg/ml glucose (X4) and 5%
   D_2O (X8); Columns: 100 cm x 4.6 mm ID (50 cm x 4.6 cm ID
   for 75Å); Pump: Altex 110A.
** V_i = V_T - V_o where V_T is the elution volume of
   D_2O. For 1000, 1400, 2000 and 3000Å columns, V_o =
   0.35 ($\pi \cdot r^2 \cdot L$). For 75Å columns, V_o was obtained
   from 2 x 10^6 dalton dextran.
*** Propagated error assuming flow rate precision of ± 1%.

Mercury porosimetry data of these packings are given in
Table IV. It is of interest to note that the pore-size distribution of CPG is significantly more narrow than that of Syn-
Chropak, a surface-modified porous silica (LiChrospher). These
different physical characteristics may help to explain the existence of micropores in SynChropak. Because of the wide pore-size
distribution of this packing, it seems reasonable that this
material also contains a population of micropores which are only
accessible to D_2O. In mercury porosimetry measurements, the
lower pore size limit is about 30Å.

TABLE IV. PHYSICAL CHARACTERISTICS OF SEC
PACKINGS FROM MERCURY POROSIMETRY

Support Pore	SynChropak (10μm diam.)			Glyceryl-CPG (37-74μm diam.)		
Diameter	100Å	1000Å	4000Å	75Å	1000Å	3000Å
Pore-size dis-	0.0044-	0.02-	0.14-	0.006-	0.09-	0.25-
tribution, μm	0.06	0.30	0.9	0.009	0.18	0.35
Dead-end volume, cc/g	1.66	1.55	0.84	0.125	0	0
V_i, cc/g*	0.92	0.96	0.82	0.33	1.35	0.89
V_o, cc/g**	1.10	1.10	1.25	0.90	1.65	1.4
Surface area, m^2/g	294	48.4	12.0	181	50	9.5

*   Pore volume
** Interstitial volume (measured to 100 psi)

Comparison of surface areas as determined by the BET and
t-curve methods (21) is another measure of microporosity since
the latter technique will estimate the surface area of pores
under 15Å in diameter. A SynChropak GPC-100 sample gave 201
m^2/g by the BET method and 216 m^2/g by the t-curve method.
The 15 m^2/g difference is attributed to micropores less than
15Å. In contrast, 75Å pore diameter Glycophase CPG was
found to have 137 m^2/g of surface area by both the BET and
t-curve methods indicating the absence of micropores.

Dead-end volume is estimated from mercury porosimetry by
measuring the amount of mercury liberated from the packing when
the applied pressure is released. This measurement approximates
the volume occupied by blind channels or pockets within the
interstitial and pore volumes. Assuming that the interstitial
volume of the bed consists totally of blind channels, then the
minimum percentage of dead-end volume within the pores of the
packing is 61 and 47%, respectively, for the 100 and 1000Å

SynChropak materials. The minimum percentage of dead-end pores
within the 4000Å SynChropak is 0%. Because of the much larger
particle diameter of the CPG packings, one would expect that
blind channels within the packed bed would be negligible. In
view of this, the 75Å CPG packing would have a maximum of 38%
of dead-end volume. The 1000 and 3000Å CPG packings have no
dead-end pores. The implication of these findings in terms of
column efficiency will be presented in a future paper (23).

Effect of Flow Rate on Elution Volume of D$_2$O and Glucose. In
order to rule out the possibility that the increased retention
volume of D$_2$O was caused by deuterium exchange on either
residual silanol groups on the packing or hydroxyl groups on the
glycerylpropylsilyl stationary phase, the elution volume of DHO
was determined as a function of flow rate. As shown in Figure 1,
there was no significant difference in elution volume when the
flow rate was varied from 0.10 to 2.0 ml/min (23.4 to 1.2 minute
residence time, respectively). For a control, the elution
volume of glucose is also given. It should be emphasized that
even if deuterium exchange were occurring, the resulting H$_2$O
molecules would not be detected. Furthermore, DHO peaks were
symmetrical; the absence of a tailed peak is further
confirmation that secondary equilibrium was not occurring.

Effect of D$_2$O Concentration on Elution Volume. If deuterium
exchange were occurring, one would also expect that the exchange
equilibrium would be dependent on D$_2$O concentration. In view
of this, 0.625 to 10% D$_2$O was injected and the resulting
retention times and peak heights are shown in Table V. The
results clearly demonstrate that there was no D$_2$O concentration dependency of either retention volume or peak height.

TABLE V. EFFECT OF INJECTION CONCENTRATION
ON PEAK HEIGHT AND RETENTION VOLUME OF D$_2$O*

D$_2$O Concentration, %	Vr, ml**	Height, cm**	DRI Attenuation
10	2.55	14.1	16
5	2.52	14.2	8
2.5	2.54	14.3	4
1.25	2.54	14.1	2
0.625	2.54	14.2	1

*   Chromatographic conditions: See Table II, 100Å column
** Average of triplicate 20µl injections

6M Urea as the Mobile Phase. The only possible partitioning mechanism that could be responsible for D_2O retention is hydrogen bonding to the glycerylpropylsilyl stationary phase which is highly unlikely because of competition between D_2O and the H_2O mobile phase. However, to rule this out, D_2O and glucose were chromatographed in a 6M urea mobile phase using a 100Å column. The results, given in Table VI, are similar to the data obtained using water as the mobile phase (Table II), indicating that the urea mobile phase had no significant effect on elution volume of D_2O.

It is of importance to note that it was difficult to prepare a 5% D_2O solution in 6M urea so that the concentration of urea would be identical to that of the mobile phase. Because of the high urea content, a relatively small difference between the urea concentration in the injected solution and in the mobile phase, produced a urea peak. In view of this, the urea content of the injected solution was adjusted to minimize interference.

TABLE VI.   ELUTION OF D_2O IN 6M UREA*

D_2O, V_r (ml)	2.59
Glucose, V_r (ml)	2.32
Δ, ml	0.27
V_i, ml	1.38
Micropore volume, %	19.3 ± 1.7

*   Chromatographic conditions: Flow: 1.0 ml/min; Chart speed: 2.5 in/min; 100Å Synchropak column. See Table II for other conditions.

Effect of Temperature on Elution Volume. The heat of solution of a solute (ΔH) (heat loss when 1 mole of solute is transferred from the mobile phase to the stationary phase) is related to the partition coefficient (K) as follows:

$$\text{Log } K = \frac{-\Delta H}{2.30 \ RT} + C \tag{2}$$

Since $K = k' \ V_M/V_S$ where k' is the capacity factor $[k' = (t_r - t_0)/t_0]$, t_r and t_0 are the elution times of a retained and unretained peak, respectively, V_M is the volume of mobile phase, V_S is the volume of stationary phase and C is a constant, then

$$\text{Log } k' = \frac{-\Delta H}{2.30 \ RT} + C' \tag{3}$$

Thus, ΔH can be readily determined by plotting log k'
versus 1/T. If ΔH is zero, there are no solute-packing interactions other than an entropic contribution (size separation).
Since, by definition, k' $\geq$ 1, the retention time of glucose
was used for t_o and the retention time of D_2O was used for t_r.
The retention times of glucose and D_2O as a function of column
temperature using a 300Å SynChropak column are in Table VII.
As indicated, the percent difference in retention time between
D_2O and glucose was about 4.5% for all temperatures. These
results were close to the 5.2% difference obtained from Table II.
The smaller value obtained in this study was probably caused by
differences in the two lots of silica used in the colums.

TABLE VII.   EFFECT OF COLUMN TEMPERATURE ON THE ELUTION
TIME OF D_2O AND GLUCOSE USING A 300Å SYNCHROPAK COLUMN*

		tr, min			
Column Temp, °C	Pressure, psi	Glucose	D_2O	Difference, %	k', D_2O
29	420	6.512	6.802	4.4	0.0445
39	348	6.496	6.776	4.3	0.0431
49	290	6.468	6.752	4.4	0.0439
60	246	6.436	6.732	4.6	0.0459
70	218	6.422	6.712	4.5	0.0451

*   Chromatographic conditions: Mobile phase: H_2O; Flow: 0.5
ml/min; Chart Speed: 5 cm/min; Volume injected: 20µl;
Sample concentrations: 1.3 mg/ml glucose and 5% D_2O;
Detector: RI X8; Column: 25cm x 4.6mm ID SynChropak 300Å.

The decrease in solute retention time with column temperature was caused in part by the expansion of mobile phase as it
entered the heated column. For example, there was a 1.3-1.4%
increase in flow rate when the temperature was increased from 29
to 70°C. The predicted value based on the expansion coefficient
of water is 0.8%.
As shown in Table VII there appears to be no significant
change of k' with respect to temperature. These data were
plotted using Equation 3 and from linear regression analysis,
the heat of solution was +0.18 Kcal/mole. Since ΔH should be
negative, this low value is obviously caused by experimental
error. Furthermore, the ΔH calculated from the standard error
of the estimate ($\pm$1 standard deviation units) of the linear
regression line is $\pm$0.17 Kcal/mole. Since ΔH is zero or is
very close to zero, Equation 3 reduces to

$$\log k' = C' \tag{4}$$

and the free energy change when DHO is transferred from the
mobile phase to the stationary phase is of the form G=TΔS.
Thus the retention time of D_2O is caused by entropic rather
than enthalpic interactions with the packing. These results
confirm that the existence of micropores must be responsible for
the difference in elution volume between glucose and D_2O.
     The effect of temperature on column efficiency is also shown
in Figure 2. As expected, the number of theoretical plates
generated by D_2O was significantly greater than for glucose
because of its higher diffusion coefficient. The temperature
dependency of glucose appears to be significantly greater than
for D_2O. For example, a column temperature change from 29 to
70°C, results in a 50% increase in efficiency for glucose as
compared to only 10% for D_2O. Since the relationship between
temperature and diffusion coefficient is linear as predicted by
the Wilke-Chang equation, one would expect a much higher plate
count for D_2O. A possible explanation for these relatively
low values for D_2O could be disruption of the packed column
bed at elevated temperatures which would affect the narrower
D_2O peak more than the glucose peak.

Conclusions

From these studies with SynChropak SEC packings and controlled
porosity glass, it is concluded that the silica packing contains
a population of micropores which are differentially accessible
to low molecular weight probes of total permeation volume. It
is not known, however, if the microporosity in the 100 and 300Å
SynChropak SEC packings is the result of the rather wide pore-
size distribution and whether all silicas contain micropores.
     The existence of micropores in a SEC packing and the
fractionation of low molecular weight probes presents a dilemma
as to what should be used as V_T in calculating K_D of high
molecular weight species. It is recommended that the corresponding monomer (except in the case of proteins) be used when
constructing a calibration curve for a given polymer. For
example, in the case of cellulosics, glucose would be the low
molecular weight calibrant of choice. D_2O is best used to
determine column efficiency because of its sensitivity toward
chromatographic peak broadening and extracolumn effects ([23]).
However D_2O may still be used to estimate V_T in some cases.
     In view of Freeman's studies on the use of normal alkanes
and polystyrenes to probe the macroporosity of porous materials
([24]), the results presented here would suggest that low molecular weight species ranging from twenty (deuterium oxide) to
several thousand daltons may be used to define microporosity of
a SEC support. The ease with which this is achieved may allow
routine examination of microporosity in new support materials
and a more exact definition of total permeation volume in SEC.

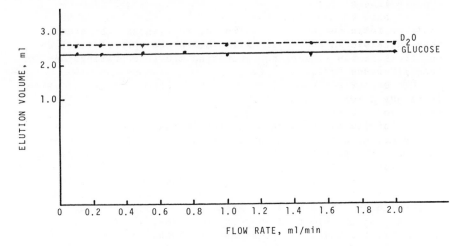

Figure 1. Influence of flow rate on elution volume of D_2O and glucose. The column was a SynChropak 100Å column. See Table II for conditions.

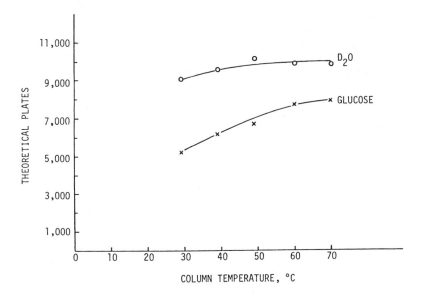

Figure 2. Influence of temperature on column efficiency using a SynChropak 300Å column. See Table VII for conditions.

Acknowledgments

The helpful discussions with Walter J. Freeman and the excellent technical assistance of David Allen Smith are appreciated. We also thank James F. Carre for providing and interpreting the porosimetry and BET data.

Literature Cited

1. Bio-Rad Laboratories "A Laboratory Manual on Gel Chromatography"; Richmond, CA, 1971.
2. Karch, K.; Sebestion, I.; Halasz, I.; Engelhardt, H. J. Chromatogr. 1976, 122, 171.
3. Rochas, C.; Domard, A.; Rinaudo, M. Eur. Polym. J. 1980, 16, 135.
4. Marsden, N.V.B. Ann. N. Y. Acad. Sci. 1965, 125, 428.
5. Yoza, N.; Ohashi, S. J. Chromatogr. 1969, 41, 429.
6. Ohashi, S.; Yoza, N. J. Chromatogr. 1966, 24, 300.
7. Obrink, B.; Laurent, T.C.; Rigler, R. J. Chromatogr. 1967, 31, 48.
8. Marsden, N.V.B. J. Chromatogr. 1971, 58, 304.
9. Barth, H.G.; Regnier, F.E. J. Chromatogr. 1980, 192, 275.
10. Barth, H.G. J. Liq. Chromatogr. 1980, 3, 1481.
11. Barth, H.G.; Smith, D.A. J. Chromatogr. 1981, 206, 410.
12. Neidhart, B.; Kringe, K.P.; Brockmann, W. J. Liq. Chromatogr. 1981, 4, 1875.
13. Grushka, E.; Colin, H.; Guiochon, G. J. Liq. Chromatogr. 1982, 5, 1391.
14. Neidhart, B.; Kringe, K.P.; Brockmann, W. J. Liq. Chromatogr. 1982, 5, 1395.
15. McCormick, R.M.; Karger, B.L. Anal. Chem. 1980, 52, 2249.
16. Berendsen, G.E.; Schoenmakers, P.J.; Galen L.D.; Vigh, G.; Puchory, Z.V.; Inczecly, J. J. Liq. Chromatogr. 1980, 3, 1669.
17. Slaats, E.H.; Markovski, W.; Fekete, J.; Poppe, H. J. Chromatogr. 1981, 207, 299.
18. Kristulovic, A.M.; Colin, H.; Guichon, G. Anal. Chem. 1982, 54, 2438.
19. Billet, H.A.H.; van Dalen, J.P.J.; Schoenmakers, P.J.; Galan, L.D. Anal. Chem. 1983, 55, 847.
20. Snyder, L.R.; Kirkland, J.J. "Introduction to Modern Liquid Chromatography"; J. Wiley and Sons: New York, 1979; p. 207.
21. Lippens, B.C.; Linsen, B.G.; de Boer, J.H. J. Catalysts 1964, 3, 32.
22. Unger, K.K. "Porous Silica"; Elsevier Scientific Publishing Co.: Amsterdam, 1979.
23. Barth, H.G., results to be published.
24. Freeman, D.H.; Poinescu, I.C. Anal. Chem. 1977, 49, 1183.

RECEIVED December 20, 1983

Methylene Chloride–Hexafluoroisopropyl Alcohol (70/30)

Use in High-Performance Gel Permeation Chromatography of Poly(ethylene terephthalate)

JAMES R. OVERTON and HORACE L. BROWNING, JR.

Research Laboratories, Eastman Chemicals Division, Eastman Kodak Company, Kingsport, TN 37662

The solvent system 70/30 methylene chloride/ hexafluoroisopropanol has been in use in our laboratory since 1977 as a solvent for poly(ethylene terephthalate) (PET) and other semicrystalline polar polymers. Some advantages of this solvent are: it provides rapid room temperature solubilization; it is transparent at 254 nm (U.V.); it is a solvent for polystyrene; and it is a minimum boiling azeotrope. Disadvantages are its low boiling point (36°C) and the potential safety hazard it represents. The combination of appropriate HPGPC equipment and this solvent system reveals heretofore unrecognized features of the molecular weight distributions of polyesters.

Poly(ethylene terephthalate) (PET) has been analyzed by gel permeation chromatography (GPC) routinely for many years.([1-7]) During this time, satisfactory results have been obtained with several solvent systems, the most common being m-cresol. The high viscosity of m-cresol requires that it be used at elevated temperatures, and the associated handling difficulty is sufficient reason for finding a replacement. This paper will present some of our experience with the solvent system 70/30 (v/v) methylene chloride (MeCl$_2$)/hexafluoro-isopropanol (HFIP). Some comments regarding the use of m-cresol are included.

Solvent System Properties

The ratio of 70/30 (v/v) $MeCl_2$/HFIP was chosen because it is a
minimum-boiling (37°C) azeotropic mixture. The exact
composition can be reproduced by distillation from a mixture of
approximately the correct ratio, and one can easily reclaim
>90% of the solvent used by simple distillation. In view of
the cost of HFIP the ability to reclaim solvent is an important
consideration.

In a kinetic sense, the system is a better solvent than HFIP
alone. We postulate that $MeCl_2$ swells the amorphous regions
of PET thereby providing HFIP with an easy access to the
crystalline regions. This swelling action does not occur with
HFIP alone, and the dissolution process takes much longer. At
room temperature, amorphous PET is instantaneously solubilized
by this solvent system. PET that has been annealed for >24 hr
at 220°C to yield maximum crystallinity dissolves in <4 hr at
room temperature. PET annealed in this manner does not dissolve
in pure HFIP after 14 days at room temperature. Poly(butylene
terephthalate) and aliphatic polyamides are soluble in this
solvent system. Polystyrene is also soluble, which permits
conventional calibration and the use of the universal
calibration approach. We have determined the Mark-Houwink
relationships for PET and polystyrene in 70/30 $MeCl_2$/HFIP to be

$$\{\eta\}_{PET} = 4.034 \times 10^{-4} \overline{M}_w^{0.691}$$

$$\{\eta\}_{PSTY} = 7.998 \times 10^{-4} \overline{M}_w^{0.54}$$

where $\{\eta\}$ is the inherent viscosity determined at 0.5 g/dl and
25°C.

The solvent system, which is transparent at 254 nm, permits
the use of a UV detector system. This is a distinct advantage
for high performance GPC where low sample loadings are necessary
and refractive index detectors may provide only marginal
sensitivity.

There are two disadvantages with this solvent system.
First, the low boiling point (37°C) can lead to handling
difficulty. We found it necessary to replace the Waters 6000A
pump in the Waters Model 244 high performance liquid
chromatograph (HPLC) with a Waters M45 pump to avoid an
occasional interruption in flow which we assured to be caused by
vapor lock. Second, there are health hazards associated with
the use of HFIP, and hygenic laboratory procedures should be
followed. The system should not be used prior to consulting the
HFIP Product Information and Material Safety Data Sheet from
Du Pont.

Experimental

This work was done with a Waters Model 244 liquid chromatograph
having two Du Pont Bimodal IIS columns (29,000 plates/meter) and
a Linear dual-pen recorder. Also used was a Waters Model 440 UV
absorbance detector. Samples were run at 0.1% (w/v) using an
injection volume of 25-μL and a flow rate of 1 mL/min. The
system was calibrated with polystyrene standards from Pressure
Chemical Co. according to the universal calibaration procedure.
Data collection and computation were done with an Intel 80/30
microprocessor.

Results

A typical GPC curve for PET prepared by melt-phase
polymerization is shown in Figure 1. The small peak on the low
molecular weight side of the distribution is caused by the
cyclic trimer of PET that is present at ∿1.5 wt % in
melt-phase polymer.(8) A sample prepared by solid-phase
polymerization is shown in Figure 2. This sample has a higher
molecular weight than the melt phase sample, and the presence of
cyclic tetramer and cyclic dimer, as well as the cyclic trimer,
can be distinguished. The identities of these peaks were
verified by spiking the samples with knowns. These features
cannot be seen on chromatograms run on Styragel columns in
m-cresol at 100°C because of inadequate resolution. Such a
curve is shown in Figure 3. Solid-phase polymerization of PET
reduces the cyclic oligomer content. This is because the
polymer crystallizes to ∿50% during polymerization, with
cyclics being excluded from the crystalline phase. The
thermodynamic equilibrium concentration is then reestablished in
the amorphous regions (during solid phase build up), and
therefore, based on whole polymer, there is approximately a 50%
reduction in cyclic oligomer content. Since cyclic oligomers
(cyclic trimer) are known to cause processing difficulties, the
determination of the cyclic trimer content of PET is often
desirable. By monitoring the 0-2 v integrator output of the
detector with the second pen of the dual-pen recorder, we can
simultaneously generate a chromatogram at two sensitivities.
This is illustrated in Figures 4 and 5. By taking the ratio of
the cyclic trimer peak from the high-gain signal to the polymer
peak from the normal signal, the cyclic trimer content of the
sample can be calculated. The values calculated from Figures 4
and 5 are 1.5% and 0.8%, respectively.
The Waters Model 244 liquid chromatograph is not equipped
with a thermostated oven and, therefore, operates at ambient
temperature. We have observed some variations in flow rate due
to laboratory temperature changes. Flow rate variations can be
illustrated by comparing the cyclic trimer elution volume in

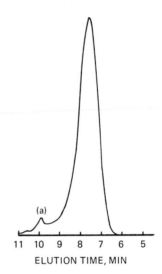

Figure 1. Melt-phase
PET (a) cyclic trimer.

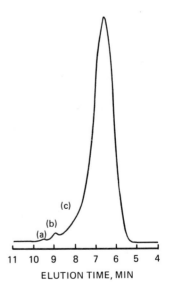

Figure 2. Solid-phase
prepared PET, (a) cyclic
dimer, (b) cyclic trimer,
and (c) cyclic tetramer.

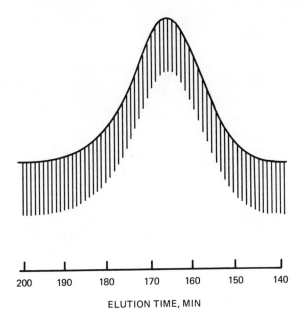

ELUTION TIME, MIN

Figure 3. Melt-phase PET, m-Cresol, 100 °C.

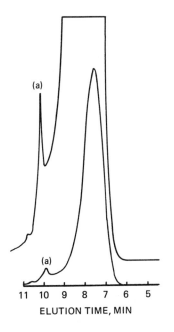

ELUTION TIME, MIN

Figure 4. Melt-phase PET run at two gains simultaneously (a) cyclic trimer.

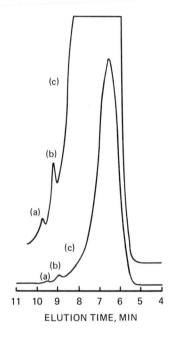

Figure 5. Solid-phase prepared PET, (a) cyclic dimer, (b) cyclic trimer, and (c) cyclic tetramer.

Figure 4 to that in Figure 5. Our computer software has been
modified to allow for these changes by using the elution time of
cyclic trimer as a measure of flow rate for a given run. The
system is calibrated by using polystyrene spiked with cyclic
trimer. For each run, elution volumes are normalized on the
basis of cyclic trimer elution. This technique assumes constant
flow rate during each run and compensates for run-to-run
variations.

Distributions to date yield values of $\overline{M}_w/\overline{M}_n > 2.0$. The
theoretical value of $\overline{M}_w/\overline{M}_n$ and the often-quoted experimental
value of 2.0 are only for linear species.(9) Consider the
effect of 1.5% cyclic trimer (ignoring the low concentration of
other cyclics) on the value of $\overline{M}_w/\overline{M}_n$. For $\overline{M}_w$ = 40,000,
$\overline{M}_n$ (linear) = 20,000. The presence of 1.5% cyclic trimer
(M=576) lowers M_n to 13,000 with essentially no effect on $\overline{M}_w$
and $\therefore \overline{M}_w/\overline{M}_n \simeq 3$. Because of the compact structure
of the cyclic trimer it elutes later than the linear species of
equivalent mass. The perceived mass of cyclic trimer by the GPC
column is actually ~ 275. In the example cited, the presence
of 1.5% of mass 275 lowers M_n to about 10,000 and $\therefore \overline{M}_w/\overline{M}_n$
$\simeq 4$.

Other workers have suggested that in a polar solvent such as
m-cresol or hexafluoroisopropanol, PET will undergo rapid ester
interchange leading to the "equilibrium distribution" having a
ratio of $\overline{M}_w/\overline{M}_n$=2.0.(6, 7) These workers failed to recognize
that the equilibrium distribution in a dilute solution is not
the same as equilibrium distribution in the absence of a
diluent.(10, 11, 12) In dilute solution, intramolecular ester
interchange dominates, and the equilibrium distribution consists
mostly of cyclic species. In our laboratory, we have been able
to show under conditions where ester interchange does occur in
solution that at a concentration of 1% polymer (w/v) the
equilibrium distribution contains >75% cyclic trimer. The
result of ester interchange in solution is, therefore, to
broaden the distribution by the generation of cyclic species.

Conclusions

The azeotrope 70/30 MeCl$_2$/HFIP is an excellent solvent for PET
and similar polymers, as well as for polystyrene. This
combination, along with its UV transparency, makes it an
excellent GPC solvent. The Du Pont Product Information and
Material Safety Data Sheet on HFIP should be consulted before
using this system.

Literature Cited

1. J. R. Overton, J. Rash, and L. D. Moore, Jr., Sixth International GPC Seminar Proceedings, Miami Beach, Florida October 7-8, 1968, p. 422.
2. G. Shaw, Seventh International GPC Seminar Proceedings, Monte Carlo, 1969, p. 309.
3. L. D. Moore, Jr., and J. R. Overton, J. Chromatogr., 55, 137 (1971).
4. Y. Ishida and K. Kawai, Shirnadzu Hyoron, 29112, 89 (1972).
5. J. R. Overton and S. K. Haynes J. Polym. Sci. Part C, 43 9 (1973).
6. E. E. Paschke, B. A. Bidlingmeyer, and J. G. Bergmann, J. Polym. Sci. Polym. Chem., 15 983 (1977).
7. M. Sang, N. Jin, and E. F. Jiang, J. Liq. Chromatog., 5 (9), 1665 (1982).
8. S. Jabarin and D. C. Balduff, J. Liq. Chromatog., 5 (10), 1825 (1982)
9. P. J. Flory, J. Chem. Phys., 12, 425 (1944).
10. H. L. Browning, Jr. and J. R. Overton, Polymer Prepr., 18 237 (1977).
11. H. Jacobson and W. H. Stockmayer, J. Chem. Phys. 87, 931, (1965).
12. H. Jacobson, C. D. Beckmann, and W. H. Stockmayer, J. Chem. Phys., 18, 1607 (1950).

RECEIVED October 20, 1983

Shear Degradation of Very High Molecular Weight Polymers in Gel Permeation Chromatography

D. McINTYRE, A. L. SHIH, J. SAVOCA, R. SEEGER, and A. MacARTHUR

Institute of Polymer Science, The University of Akron, Akron, OH 44325

The degradation of very high molecular polymers in
GPC is demonstrated to occur in the gel columns, to
begin at a critical molecular weight depending on
the polymer structure, and to follow a power law dependence on MW after the onset of degradation. A
loop model of entanglement is advanced to explain
the degradation, and guidelines to minimize degradation are explicitly described.

An earlier experiment in these laboratories reported that very
high molecular weight polystyrene (PS) was degraded in gel permeation chromatography (GPC) columns operating at relatively low
pressures (125 psi) and low elution rates (1ml/min) (1). The degraded very high molecular weight polystyrene (MW 44×10^6) was recovered from the eluent, and its molecular weight was determined
by intrinsic viscosity measurements. The molecular weight of the
original polymer, 44×10^6, had been decreased to 19×10^6. Thus the
original polymer chain had on the average been cut to less than
one-half its size in its passage through the GPC column. When the
degraded molecular weight was used as the correct molecular weight,
the degraded polymer nearly fit the GPC calibration curve of elution volume-molecular weight that had been established with much
lower molecular weight polystyrenes. Since earlier work (2) had
shown that a 10×10^6 MW polystyrene did obey the GPC calibration
curve, the onset of measurable degradation had to occur at a molecular weight greater than 10×10^6.

It seemed worthwhile to explore the generality of the earlier
finding of chain degradation in PS at very high molecular weights,
since the degradation only had been shown to occur with polystyrene in a given set of columns, using a conventional mechanical
configuration, while operating at a low shear rate (or equivalently elution rate). Consequently, both the physical set-up of the
GPC columns and the chemical structure of the chromatographically
separated polymers were varied in this study. High molecular

weight polydimethylsiloxane (PDMS) and PS over a range of molecular weights were examined. Benzene was used as a solvent.

The flow rate and mechanical constrictions in the tubing were
varied while attempting to measure degradation in the GPC. The
change in flow rate is related to the pressure drop and therefore
to the shear rate in the columns. The operating pressure was
varied only over a narrow range (50 psi to 150 psi, or an equivalent flow rate of 1ml/min to 0.25 ml/min). Severe constrictions
to the flow of liquids in the column occur in the 10 μm fritted
filter at both the inlet and the outlet of each packed column and
also in the interstices of the packing in the column. Either of
these constrictions might be the source of the shearing stresses
for polymer degradation. Since a 44 million MW polystyrene has
an unperturbed radius of gyration of 0.25 micron[3] and therefore
would have some instantaneous chain segment end-to-end distances
that would approach the size of some of the pores in the fritted
filter, the effect of the filter on the degradation was carefully
examined first.

PDMS was chosen to determine if polymers other than polystyrene degrade during GPC analyses, and, if so, at what molecular
weights. PDMS was chosen because it is an even more flexible
chain and also has a large chemical difference in the chain backbone structure. Although the exact relation between chain flexibility, chain entanglements, and shear degradation is not well
understood, these experiments use dilute polymer solutions so that
the entanglements ought to be related to the characteristic parameter (or relative unperturbed size) of the single polymer chain.
Consequently the degradation of high molecular weight PDMS in GPC
columns ought to be different from the degradation of the less
flexible and purely hydrocarbon backbone of PS. Also, it was felt
that the PDMS backbone rupture would not involve a free radical
mechanism and subsequent chain transfer reactions. These findings are particularly timely now because there has recently been
speculation that there is extensive degradation of all polymer
chains in the newer and faster, high-pressure GPC instruments[3,4].
Other polymers with a greater range of flexibility were also
studied.

Experimental

Polymers - The PS, PDMS, polyhexylisocyanate (PHIC), and polyisoprene (PI) samples had been extensively characterized to determine
molecular weights, molecular sizes, and thermodynamic parameters
(5, 6, 7). The samples were anionically polymerized using butyl
lithium as the initiator. The pertinent data are shown in Table I.
Polyisobutylene/PIB polymers were obtained by fractionation of
commercial polymers and their molecular weights were measured (8).

Solvents. Benzene - Baker, reagent grade; Cyclohexane - Matheson,
Coleman and Bell (MCB), reagent grade; Tetrahydrofuran - Fisher
Scientific, reagent grade.

Table I. Identification and Molecular Weight of Polymers

Polymer		M_W	Source
PS	13	4.4×10^7	Ref. 5
	18	2.72×10^7	
	11	9.6×10^6	
	9	4.5×10^6	
	25166	4.11×10^5	
	61970	2.6×10^6	Waters Associates
	25167	8.67×10^5	
	41995	9.82×10^4	
PIB	B	1.5×10^6	Ref. 8
	E	1.2×10^6	
	F	6.5×10^5	
	PIIA	1.5×10^5	
PDMS	5-1	2.0×10^7	Ref. 6
	5	1.2×10^7	
	A	6.8×10^6	
	B	4.4×10^6	
	A-1	1.46×10^6	
	A-2	5.5×10^6	
	A-3	5.5×10^5	
PHIC	11	4.24×10^4	
	22	5.8×10^4	Ref. 7
	33	1.33×10^5	
	44	2.30×10^5	
	66	1.31×10^6	
PI	2E7	7.2×10^6	Ref. 7
	20M	1.8×10^6	
	7E5	7.6×10^5	

GPC Instrument Operation

1. High Molecular Weight Polymers in Routine Degradation Experiments. Waters Associates Ana-Prep and 501 GPC were used for separation of high molecular weight PS, PDMS, PI, and PIB fractions. Five four-foot Styragel columns were connected in the following sequences (Set A) using a differential refractometer as the detector.

Set A

one: 7×10^5 to 5×10^6 Å
one: 7×10^5 to 5×10^6 Å
one: 7×10^5 to 5×10^6 Å
one: 1.5×10^5 to 7×10^5 Å
one: 5×10^4 to 1.5×10^5 Å

The size designations are those given by Waters Associates. This set had a plate count of 680 PPF when o-dichlorobenzene was the solute. Samples were prepared on a weight-to-volume basis. Each sample was run at several different concentrations in the range of 0.05 - 0.2 g/dl in order to extrapolate the peak position to zero concentration. Full loop injections were used for all solutions. A 2.5 ml siphon was used at the elution end.

PS 13 and PS 18 were also run through Set A at a reduced flow rate of 0.5 ml/min and reduced concentration. No significant changes occurred in the peak position and in the shapes of the curves.

2. High Molecular Weight Polymers in Cyclohexane and also in Special Column Arrangements. Waters Associates Ana-Prep and 501 GPC were used. One four-foot Styragel column of 5×10^6 pore size was connected to a pump and a differential refractometer detector to determine the effect of fritted discs on degradation.

Single columns of different pore size were used to determine the effect of gel pore size on degradation.

Single columns were used to determine the effect of solvent power on degradation.

Samples were prepared on a weight-to-volume basis. Full loop injections were used for all solutions, and polymer from the GPC eluent was recovered for characterization by taking all eluent solution 2 counts before and 2 counts after the polymer elution peak.

Viscosity Measurements. A Zimm-Couette type low shear viscometer was used. The intrinsic viscosities were estimated from single concentration viscosity measurements using the equations for the concentration dependence of the specific viscosity (5,6). The Mark-Houwink equation was used to determine M_V (5,6).

Experimental Design

a) Measurement of Degradation. The experiments were carried out to elucidate the roles of both physical and chemical variables in

the GPC degradation of high molecular weight polymers to lower
molecular weight polymers. Therefore, a measure of degradation
had to be chosen that was independent of GPC. Although viscosity,
light scattering, and sedimentation measurements of molelcular
weight have been made, only the viscosity measurements are reported here. Although the whole molecular weight distribution is
desirable for analysis, only the single viscosity - average moment
of the molecular weight distribution was determined. A simple
measurement of degradation was determined as:

$$\% \text{ Degradation} = \%D = (100 - \% \text{ Decrease MW}) = 100 \left[1 - \frac{(MW) \text{ after GPC}}{(MW) \text{ before GPC}}\right]$$

b) <u>Physical Variables</u>. The effect of shear rate on degradation
was evaluated by changing flow rates, pore size, packing geometry,
column length, solution viscosity, and frits in the columns.

c) <u>Chemical Variables</u>. The effect of the backbone bond strengths
and the flexibility of the polymeric chain was evaluated by studying the degradation of polymers of different backbone structures
[$\{C-C\}$, $\{Si-O\}$], of flexible polymers with different chain flex-
ibilities at constant backbone structure [PIB, PS], and of rigid
polymers [PHIC].

d) <u>Physico-Chemical Effects</u>. Polymer concentrations were kept
low in order to reduce the solution viscosities and measure only
the effect of the GPC on single polymer chains. At the highest
MW's the concentrations were always <0.02%. Both poor and good
solvents were used to decrease solution viscosities and in some
cases enhance adsorption of the polymers to the packing.

<u>Results and Discussion</u>

<u>Physical Variables</u>. Very dilute solutions of the PS and PDMS in a
syringe were pushed through a 10 μm fritted filter similar to the
filter in the GPC columns. The relative viscosities of the filtered and unfiltered (original) solutions were determined in a low
shear viscometer. The values of the relative viscosity are given
in Table II. The data indicate that there is no degradation within the experimental error of ±2%. In separate experiments a single Styragel column without the fritted filter at the inlet end was
used to analyze the polymer in the conventional manner, but the
eluent containing the polymer was collected and analyzed. The
measured relative viscosities and the estimated intrinsic viscosities and molecular weights are given in Table III. Both the PS
and PDMS of similar molecular weight have comparable degradation
in the single GPC column. Degradation occurred even when the flow
rate was reduced from 1 ml/min to 0.25 ml/min.
    It is clear that the packed Styragel column is responsible
for the degradation of the polymer. In all cases the pores of the
packing are sufficiently large to accommodate portions of the

polymer chain, since the flow time of the PS and PDMS is longer
than the time required for plug flow of solvent. Since the 10×10^6
MW PS sample falls on the linear portion of the log MW - elution
volume calibration curve, the degradation at even this high molecular weight cannot be greater than a few percent - which would
be an acceptable level in many polymer characterization studies.
Similarly the 7×10^6 MW PDMS sample falls on its empirically established GPC calibration curve using low molecular weight samples.
Since the radii of gyration of the PS and PDMS molecules used in
these experiments are of comparable magnitude [3460Å)(PS), 2750Å
(PDMS)], and since the molecular weights of the repeat units [104
(PS), 74 (PDMS)] are not too different, it appears that a molecular weight of 10×10^6 or a radius of gyration of 2000Å represents
a limit for the reliable estimate of the molecular weight by GPC
of a flexible polymer chain with stable main chain bonds like C-C
or Si-O.

Table II. Relative Viscosities and (η_{rel}) and Reduced Specific
          Viscosities ($\eta_{sp/c}$) of Polymer Solutions Passed
          Through Fritted Filter (10μ)

Polymer (MW)	Conc.(g/dl)	η_{rel}	η_{sp}/c(dl/g)	$<S^2>^{\frac{1}{2}}$ (3460Å)
PS (27×10^6)				
unfiltered	0.024	2.159	47.3	
filtered	0.024	2.130	48.3	
PDMS (20×10^6)				(2750Å)
unfiltered	0.030	1.631	21.0	
filtered	0.030	1.629	20.9	

Table III. Measured Relative and Reduced Viscosities. Estimated
           Molecular Weight, Intrinsic Viscosities of Polymers
           Passed Through GPC Columns in Benzene

Polymer (MW)	C(g/dl)	η_{rel}	η_{sp}/C	[η]dl/g	M_v	%degradation
PS (27)	0.0074	1.182	24.6	23.1	17×10^6	33 (see note)*
PDMS (20)	0.0077	1.111	14.4	13.8	14×10^6	30

*in THF, degradation of PS was also 33%

These experiments were made using benzene as a solvent. The
earlier experiments (1) were made using THF, with an added antioxidant, as a solvent. Other experiments (2) had shown that THF
without an antioxidant induced drastic degradation of PS solutions.
    Earlier data on the GPC degradation of 27×10^6 MW PS in THF
were compared with the current data in benzene. The degradation

in both solvents is comparable. Consequently the degradation cannot be attributed to a solvent-induced degradation. This observation is reinforced by the new observation that both PDMS and PS
have comparable chain degradation. If the degradation were to
come from radicals generated in the solvent, then PDMS would degrade much less than PS.

PHIC was chromatographed with the same columns and solvents
as PS and PDMS. The results of this experiment are shown in Table
IV, and the comparison with earlier PS and PDMS results is summarized in the same table. The MW of PHIC did not change during
the chromatographic separation with an experimental error of ±5%.
The composition of PHIC is $\overset{O}{\underset{||}{C}}$ C$_6$H$_{13}$ and the backbone is rigid.

$$\left(C - N \longrightarrow \right)_n$$

Consequently, a relatively low molecular weight polymer (MW=1.39x
10^6) has an intrinsic viscosity (25.6 dl/g) equivalent to a PS or
PDMS more than 10x higher in MW. (Earlier GPC work on PBIC (9)
and PHIC (7) had shown no deviations from the universal calibration curve for GPC at high molecular weights). The primary role
of chain flexibility in GPC degradation rather than simple molecular hydrodynamic volume is conclusively shown by these results.

Table IV. Comparative Degradation of Stiff (PHIC) and Flexible
Chains (PS, PDMS)

PHIC	[n]dl/g	M_v(g/mol)
before GPC	25.6	1.39×10^6
after GPC	25.0 (±5%)	1.32×10^6

Comparison	MW	[n]dl/g	%D
PS	27×10^6	36	33
PDMS	20×10^6	20.5	30
PHIC	1.4×10^6	25.6	∿0(<3%)

In order to evaluate the role of chain flexibility on GPC de-
gradability when the backbone has a constant chemical structure,
namely $(C - C)$, high molecular weight PIB was chromatographed. Unfortunately the highest molecular weight fraction available had a
molecular weight of 3×10^6. Nevertheless, the degradation at the
highest MW was 25%, although there was no degradation at 1.5×10^6
MW. These data are graphically shown in Figure 1 in which all of
the data for PS, PDMS and PIB are shown. These PIB data are sup-
ported by the data of Huber and Lederer (3) on the GPC degradation
of PIB using a different GPC experimental arrangement.

An unusual backbone structure occurs in polyisoprene (PI).
The alternating single and double bonds, $(CH_2 - C(CH_3) = CH - CH_2)$, can
give rise to additional chemical enhancement of the primary degradation by free radical chain transfers and branching--even though
the flexibility of the monomer segment is similar to PIB. The
degradation is extensive in PI. For a seven-column set the

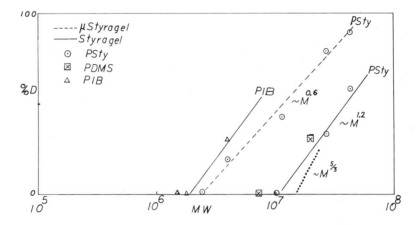

Figure 1. Per cent degradation (%D) as a function of Molecular Weight (MW) to determine onset of degradation (critical MW) and power law dependence of degradation on MW.

degradation begins below $5x10^5$ and is about 70% at $10x10^6$ MW.
These data are being rerun under various experimental arrangements
and will be reported in detail later. The significant fact is
that the degradation begins at $(2-5)x10^5$ MW and there is extensive degradation at low MW's. As a result PI can be used as a
test-probe polymer to evaluate the physical variables that affect
degradation by enhancing the primary degradation and therefore the
detection capability.

GPC degradation of polymers has been shown in four polymeric
systems using Styragel packing. Chain flexibility is an important parameter. The degradation appears to occur in the column
packing and not in the frit, although no attempt was made to
change the injection loop plumbing. Work is planned to evaluate
the effect of the shear stresses in the injection loop. The following investigations were carried out to discover those variables
in a given GPC set-up that might lessen the degradation and lead
to practical ways to minimize the degradation for routine analyses.

It is possible to alter the pore size, packing swelling, and
bead size of Styragel type packing by using different porosities,
different solvents (swelling and non-swelling), and Styragel and
µ-Styragel. Although the experiments were not exhaustive, they
attempted to sort out the variables in a rough manner. PI was
used to magnify the effect with the realization that the chemical
degradation is not simple in PI. Tables V, VI, and VII present
the scope of the survey of physical variables.

Table V. Effect of Pore Size on GPC Degradation

Polystyrene

	Expt'l conditions:		% Degradation	
	(Solvent: THF); (Styragel)	MW:	$44x10^6$	$27x10^6$
a)	6-col., large pores (no 10^3 col.)		58	33
b)	1-col., 10^3 por.		86	78

Polyisoprene

	Expt'l conditions:		% Degradation	
	(Solvent: THF); (Styragel)	MW:	$10x10^6$	$3x10^6$
a)	1-col (10^7)		18	15
b)	1-col (10^3)		37	21

From Table V it is an inescapable fact that a small pore enhances the degradation. For PS the 7-column set had no 10 pore-
size column, and yet the whole extra battery of 6 higher pore-size
columns degraded PS less than the one low pore-size column. The
PI results are similarly illustrative. When just one low

pore-size column was used the degradation is twice that of the degradation with a large pore-size column.

Table VI.   Effect of Gel Swellant on GPC Degradation

Polyisoprene

Expt'l conditions:		% Degradation
Solvent:	(Styragel)	(MW: 1.8×10^6)
Cyclohexane		69
THF		36

From Table VI the influence of swelling (THF) and non-swelling (cyclohexane) solvents on the GPC degradation is clear. A non-solvent for the packing enhances the degradation almost two-fold for PI. Obviously the better the gel swellant, the more likely the lessening of degradation. An explanation of this effect can only be reconciled to be the combination of two related factors. First, the non-swellant increases the surface adsorption of a chemically similar polymer solute and gel. Second, the bead pore will have a lower overall size in the non-swellant. Both of these factors would lead to increased degradation as discussed later, although a change of the texture of the pore surface might alter the dependence on the average pore size alone.

Finally Table VII demonstrates the effect of the bead size on the GPC degradation of polymers. The small bead size (10μ vs 60μ) would decrease the interstitial volume. Unfortunately, the increased pressure also increases the shear rate so that both decreased interstitial volume and high shear rates occur simultaneously. However, it is clear that both the MW at the onset of degradation and the amount of degradation are higher. From an analytical chemist's viewpoint, lower pressures and larger interstitial volumes are to be preferred if the goal is to decrease systemic errors in the GPC analysis of high MW polymers. It is interesting to compare the results of Rooney and VerStrate (4) in which polystyrene of 4×10^6 MW at a flow rate of 1ml/min showed a 40% degradation @ 135° in TCB with Showdex 800. The 22% degradation in Table VII, run at 1ml/min but with μ-Styragel at 40°C, would be expected to be smaller than that at 135°C so that the difference between 22% and 40% is not unreasonable. Rooney and VerStrate also describe a strong dependence of degradation on flow rates. In the Styragel results described earlier in this paper, there is no apparent decrease in degradation when the flow rate is lowered from 1.0 to 0.25 ml/min. The dependence of degradation on concentration was not explicitly measured in this work because the lowest detectable concentrations were used. Comparative experiments at a fixed molecular weight were carried out at approximately the same composition. There is a rapid rise of the solution viscosity of very high molecular weight polymers at low concentrations ($\sim$0.1% g/dl), called the entanglement region. Concentrations above this region must be avoided.

Table VII.   Effect of Styragel Bead Size on GPC Degradation

Polystyrene	% Degradation	
Expt'l conditions:		
MW	Styragel (R=60μ)	-Styragel (R=10μ)
Solvent: THF	(100 psi)	(1000 psi)
7-column set (Type A) 40°C		
44x10^6	58	88
27	33	78
10	0	-
8	-	22

Recommended Analytical Procedure.   A protocol to eliminate or at least minimize any systematic error in the GPC determination of high molecular weights would be:

1. Use concentrations well below the entanglement region and as low as can be detected.
2. Use the lowest possible flow speeds and pressures.
3. Use the largest interstitial volumes (large bead sizes).
4. Use swelling solvents.
5. Avoid low pore sizes.
6. Use the lowest injection speeds.
7. Avoid polymer - substrate adsorption.
8. Avoid, if possible, high temperatures or reactive solvents.

A Loop Entanglement Model Rationalization of GPC Chain Degradation.   These experimental results suggest that there are large enough shear stresses in GPC columns to break the backbone chemical bonds of polymers. The large stresses on the chain likely occur because the fast moving solvent outside the pore and the slow moving solvent inside the pore causes a velocity gradient on a portion of the large flexible chain molecule which may have appreciable portions of its chain segments both inside and outside the pore. Since the equilibrium chain segment distribution at these high molecular weights would be expected to have appreciable loop entanglements, the shear stress due to the velocity gradient would cause a locking of the loops on the time scale required for the repositioning of the segments either totally inside or totally outside the pore of the gel. As a result the chain backbone is broken. If there were adsorption of segments in the interstitial passages, the same mechanism would lead to degradation. If the

SIZE EXCLUSION CHROMATOGRAPHY

shear field were sufficiently strong at any point in the apparatus, the degradation would occur whether or not there were pores
or adsorption. But the mechanism would still involve a loop entanglement to concentrate the stress, unless the stress were given
as a very short time impulse in which very short chain segments
could not respond immediately. Consequently, it is conjectured
that most flexible polymers will be degraded in GPC experiments
if the polymer size is sufficiently large to allow considerable
loop formation in the distribution of the chain segments.

The loop entanglement model has been discussed elsewhere
(10-12). A brief pictorial representation of the "locking of
loops" is given in Figure 2. The entangling of loops leads to the
possible "locking of loops" when the tangle is under stress. Figure 2 illustrates three conditions for "locking". The degree of
entangling depends on the number of loops, n_ℓ. Since the number
of loops depends on the length of the chain, the number of loops
depends on a power of M. Let that power of M be $M^{5/3}$ (10,12).
As a pair-wise process the interacting of pairs to produce
"locked" loops, n^*, depends on the square of the number of loops
between two identical MW neighboring chains A and B. That is n^*
is proportional to $(n_\ell)_A (n_\ell)_B$, or $(n_\ell)^{10/3}$. In Figure 2a the
thin strand from B is locking the thick strands from A. In an adsorbed chain the "locking" is proportional to $[(n_\ell)_A(\text{adsorbed wall}$
sites$)]$ for the molecule A in the GPC column as shown in Figure
2b. For interaction with polymer loops from the gel the "locking"
is due to $[(n_\ell)_A(n_\ell)_{gel}]$ as shown in Figure 2c. Since the number
of adsorbing sites or the number of loops in the gel is a constant
for a given GPC experiment, the "locking" is proportional to
$(n_\ell^{5/3})_A$ or $M^{5/3}$.

Of course the number of "locked" loops fundamentally depends
not on the MW but on the number of statistically independent chain
segments. The loops therefore depend on the chain flexibility or
the characteristic parameter of Flory, C_∞, and the degree of polymerization (DP). Thus a stiff chain like PHIC does not easily
degrade, a flexible polymer like PS, ($C_\infty \sim 10$), degrades less than
PIB, ($C_\infty \sim 6$). There ought to be a critical size (or MW) below
which degradation in GPC does not occur. Figure 1 gives an indication of these critical values $(M)_{GPC}$, where degradation in GPC
first begins. Also the discussion of PI leads to a value of
$(M_c)_{GPC}$ in the region of $(2-5\times10^5$ g/mol. The direct relationship
between $(M_c)_{VISC}$ and $(M_c)_{GPC}$ is discussed elsewhere (12) and appears to be approximately $(M_c)^2_{VISC} \sim (M_c)_{GPC}$.

As a consequence of this model it is qualitatively easy to
anticipate when degradation will occur if $(M_c)_{VISC}$ is known. That
is, $(M_c)_{GPC}$ is $(M_c)^2_{VISC}$. A rough dependence of degradation on
$M_{exp}^{1.2}$ or $M_{theo}^{5/3}$ will not be far from the correct result for PS or
PIB if $(M_c)_{GPC}$ is estimated correctly. At high shear stresses the
$(M_c)_{GPC}$ for μ-Styragel is lower and the power law dependence in M
is lower ($\sim M^{0.6}$). A more exact description of these phenomena is
currently under investigation, theoretically and experimentally(15).

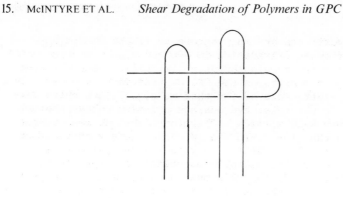

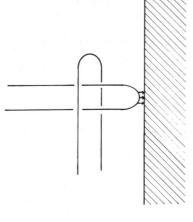

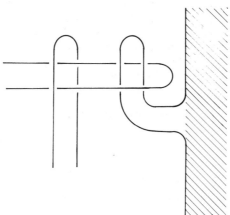

Figure 2. Loop entanglements that could lead to "locking" under stress: (a) locking of free loops; b) locking of loops by adsorption on substrates; c) locking of loops by interpenetration into loops on substrate.

Further experiments are in progress to locate the $(M_c)_{GPC}$ for the onset of chain degradation in GPC and also to determine the quantitative relationships between flexibility, entanglement, and backbone bond strengths on the GPC degradation. However, it is clear in the experiments so far that no significant chain degradation occurs in saturated hydrocarbon polymers unless the molecular weights are 5-10 million. Therefore, most MW measurements on polymers less than $(5-10) \times 10^6$ in MW can be safely carried out with a low pressure GPC apparatus, unless (1) the polymer is known from other observations to be especially susceptible to shear degradation, or (2) the MW's must be determined to accuracies better than 3%. But at very high molecular weights it is clear that the estimated GPC MW can easily be one-half or less that of the true MW. For unsaturated chains the secondary chemical reactions hasten the degradation as soon as the $(M_c)_{GPC}$ threshold is passed.

Qualitatively all of the observed GPC degradation characteristics can be rationalized by the above loop model. Reasonable estimates of the onset of degradation in GPC can be made, and estimates of the percent degradation can be made cautiously.

Literature Cited

1. Slagowski, E.L.; Fetters, L.J.; McIntyre, D.; Macromol. 1974, 7, 394.
2. McIntyre, D.; Fetters, L. J.; Slagowski, E. L.; Science 1972, 176, 1042.
3. Huber, C.; Lederer, K. H.; Polymer Letters 18, 535 (1980).
4. Rooney, J. G.; VerStrate, G.; in "Liquid Chromatography of Polymers", Cazes, J., Ed.; Dekker, New York, 1981; p. 207.
5. Slagowski, E. L.; Ph.D. Thesis, The University of Akron (1972).
6. Shih, A. L.; Ph.D. Thesis, The University of Akron (1972).
7. Kuo, C. C.; Ph.D. Thesis, The University of Akron (1980).
8. Shih, A. L.; M.S. Thesis, The University of Akron (1968).
9. Ambler, M. R.; McIntyre, D.; Polymer Letters, 13, 589 (1975).
10. MacArthur, A.; M.S. Thesis, The University of Akron (1978).
11. MacArthur, A.; Stephens, H. L.; J. Appl. Polymer Sci., 1983, 28, 1561.
12. MacArthur, A.; McIntyre, D.; Rubber Division, Toronto, May (1983), Paper #7.
13. Flory, P. J.; "Statistical Mechanics of Chain Molecules"; Wiley: New York, 1968.
14. Ferry, J. S.; "Viscoelastic Properties of Polymers" (second ed.); Wiley: New York, 1970.
15. McIntyre, D.; MacArthur, A.; Polymer Preprints, 24, August 1983, p. 102.

RECEIVED October 13, 1983

High-Efficiency Gel Permeation Chromatography
Applications for the Analysis of Oligomers and Small Molecules

A. KRISHEN

Chemical Research and Development Division, The Goodyear Tire & Rubber Company, Akron, OH 44316

High efficiency columns currently available for gel permeation chromatography of small molecules and oligomers provide high speed separations. The use of multiple detectors provides additional information which facilitates characterization and determination of the separated species.

High efficiency gel permeation chromatogrphy in the low molecular weight range is based on the same mechanism of separation as gel permeation chromatography for high molecular weight polymers. The solute components are selectively retarded according to the degree of their permeation into the solvent filled pores in the column packing. Larger molecules are excluded from the pores of the packing due to their physical size and thus elute before the smaller molecules. Beyond this similarity however, the resemblance between the two is minimal. Molecular weight distributions are the main concern for polymers while retention volumes and distinct separation of individual species are of importance in high-efficiency gel permeation chromatography in the low molecular weight range. This technique has greater similarity to both gas chromatography and high performance liquid chromatography than to gel permeation chromatography of polymers.

Gel permeation chromatography for small molecules is a relatively recent development in chromatographic techniques. In 1968 Hendrickson (1) predicted, "It appears likely that GPC for small molecules will become a new basic tool that could be called a liquid phase size spectrometer."

0097–6156/84/0245–0241$06.00/0

Advantages

In contrast to gas chromatography and other forms of
liquid chromatography, gel permeation chromatography
is applicable even when there are no differences in the
solubility, polarity, adsorption, ionic characteristics
or volatility of the molecules. Furthermore the
following three characteristics of the technique are
responsible for its increasing utility:

1. Separations are achieved with a single solvent.

2. Elution of all the components of a mixture normally
    takes place in a finite volume controlled by the
    column characteristics.

3. Additional important information on the size of the
    separated compounds can be obtained easily by com-
    parison with known compounds having similar
    characteristics; furthermore the characterization
    of the separated compounds is facilitated by the
    use of multiple detection systems.

Columns

The number of suppliers of the polymeric gels of the
polystyrene-divinylbenzene type suitable for the
analysis of small molecules has been increasing over
the years. Packed columns - usually 30 cm x 8 mm i.d.
or gels are currently available from many sources (2)
although some of them sell only the packed columns.
     Columns range in theoretical plate efficiency from
12,000 to 40,000 plates per meter as measured by
chromatographing acetone at a flow rate of 1 ml per
minute. The molecular weight range covered by each
column depends on the pore size of the gel. It is
quite common to use either a combination of columns of
different pore sizes or columns with beds of mixed
pore gels to obtain the desired range for the application at hand. A typical chromatogram obtained for a
mixture of n-alkanes is shown in Figure 1. Alkanes
of sufficient purity to serve as standards are available up to about $C_{44}H_{90}$ (M.W. 618) and low molecular
weight polystyrenes are useful as standards in the
higher molecular weight range. Most of the columns
provide sufficient resolution to separate the lower
molecular weight polystyrene standard samples into
their individual oligomeric components.

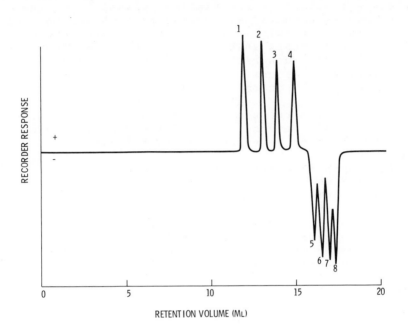

Figure 1. Gel permeation chromatogram of n-alkanes. Conditions: 610 mm x 8 mm TSKG 2000 H8 column; and tetrahydrofuran eluant at 0.5 mL/min. Key: 1, $n\text{-}C_{36}H_{74}$; 2, $n\text{-}C_{24}H_{50}$; 3, $n\text{-}C_{18}H_{38}$; 4, $n\text{-}C_{12}H_{26}$; 5, $n\text{-}C_8H_{18}$; 6, $n\text{-}C_7H_{16}$; 7, $n\text{-}C_6H_{14}$; and 8, $n\text{-}C_5H_{12}$.

Calculation of Molecular Sizes

The high efficiency columns separate individual components as distinct peaks similar to those obtained in
gas chromatography. It has been recognized that the
separation is predominantly based on the size that the
molecule exhibits in the eluting solvent under the
experimental conditions. Geometrical shape, molecular
association and solvolysis due to hydrogen bonding are
some of the factors which control the effective size
of a molecule. Hendrickson and Moore (3) and
Hendrickson (1) considered the chain length of s
molecule as the controlling factor for the elution
volume. The chain lengths of the molecules were
calculated from the bond angles and atomic radii. n-
Hydrocarbons were used as standards. Their chain
lengths can also be calculated as follows:

Chain Length (Angstroms) = 2.5+(1.25 x Number of
Carbons).
     The chain lengths of different molecules were then
compared with the chain lengths of hydrocarbons to calculate the carbon number which relates the elution volume of the compound to the elution volume of a real or
imaginary n-hydrocarbon. The "effective" carbon
number for benzene based on its elution volume was
experimentally found to be 2.85 - i.e. it eluted
near the elution volume for propane which has a
carbon number of 3.0 while the calculations based on ·
bond angles and radii of atoms indicated that benzene
would have a carbon number of 3.55. Thus corrections
to the calculations for carbon number were required.
These were derived from the experimentally observed elution behavior of various molecules.
     Smith and Kollmansberger (4) introduced the con-
cept of molar volumes as "fundamental in determining
the degree of separation." Molar volumes were calculated for a number of compounds from their density and
were expressed as ml/mol. Chang (5), Cazes and Gaskill
(6,7) and Edwards and Ng (8) provided further understanding of the basic factors involved in the separation process in gel permeation chromatography. Lambert
(9,10) combined the data from various investigators and
attempted to recalculate the results for a large number
of compounds on a common basis. The n-hydrocarbons
were used as standards and their molar volumes were
found to conform to the following equation

Molar Volume (ml/mol at 20°C) = 33.02 + 16.18 x (CAU)+
                                 0.0041 x (CAU)2

where CAU is the size of the hydrocarbon in carbon atom
units. Lambert's (9,10) summary of the factors to be
considered for predicting the elution volumes for small
molecules in tetrahydrofuran pointed out the generali-
zations.

The elution volumes for n-hydrocarbons show a
straight line relationship vs the logarithms of their
molar volumes. Molar volumes, calculated from the
densities of compounds other than n-hydrocarbons, must
be modified to have the elution volumes of these compounds conform to the same calibration line (elution
volume vs log molar volume) as that for the n-hydrocarbons. W. W. Schulz (11) related the elution behavior of branched alkanes in the range of C_7-C_{11} to
the average numbers of gauche arrangements (Zg) which
the molecule can assume.

The molecular weights of oligomeric species
corresponding to chromatographic peaks and their
identification are usually more useful than their molar
volumes or their carbon numbers. The interpretation
of experimental data can be simplified by using the
logarithm of the molecular weights rather than the
logarithms of molar volumes for standardization.
Larson (12) indicated that the elution volumes for
aliphatic hydrocarbons, aromatic hydrocarbons and
aliphatic alcohols formed three different parallel
straight lines when plotted against the logarithms of
their molecular weights. These relationships are
similar to those observed in gas chromatography for
different homologous series. The line for the
aromatic hydrocarbons was displaced higher than that
for the n-alkanes, indicating that the "effective"
molecular weight of the aromatic hydrocarbons was
smaller than their actual molecular weight while the
alcohols exhibited larger effective molecular weights
due to their hydrogen bonding with the solvent, tetrahydrofuran.

We have recently reported (13)) the use of this
technique for characterization of various compounds.
Experimental data obtained for a number of compounds
are shown in Figure 2. We calculated the "size
factors" for a number of small molecules and oligomers.
This factor is a measure of the deviation of the
elution volume of a given species from the calibration
curve for n-alkanes which is assigned a size factor
of 1. This size factor, F, is defined to be equal to
A/M, where M is the molecular weight of the compound
and A is the molecular weight of a real or hypothetical
n-alkane which will elute at the same retention volume
as the compound. Size factors for a number of

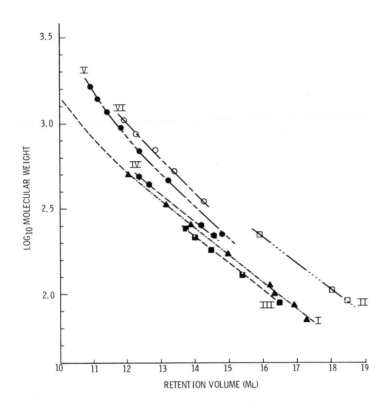

Figure 2. Molecular weight vs. retention volume relationships. Key: I, n-alkanes (C_5H_{12} to $C_{36}H_{74}$); II, toluene,
p-xylene, and diethylphthalate; III, n-alcohols (C_4H_9OH to
$C_{16}H_{33}OH$); IV, 2,6-di-tert-butyl-p-cresol, dibutyladipate,
n-didecyl phthalate, and n-didodecyl phthalate; V,
nonylphenol-formaldehyde adducts; and VI, TMDQ oligomers
(dimer to hexamer).

compounds along with their molar volumes have been
calculated (13).

Applications:

While the mechanism of retention for various compounds
can be only partially understood, the application of
the technique provides much needed information for
characterization of different types of materials.
Some of these applications, including those requiring
the use of multiple detectors, are presented here.

Phenol - Formaldehyde Resins. Oligomeric species produced by interlinking of phenolic molecules with
$-CH_2-$ moieties result in complex mixtures which cannot
be resolved by gas chromatography. The GPC system provides ample information on the individual components
for comparison of different batches of resins specially
when a dual detector system consisting of a differential refractive index and a UV absorption detector
(254 nm) is used (Figure 3). Each of the peaks observed in the chromatogram represents individual oligomers produced by addition of a monomeric unit consisting of the phenol and $-CH_2-$ since a plot of the
retention volume versus the logarithm of the molecular
weight of the oligomers produces a straight line
(Figure 1; V).

2,2,3-Trimethyl-1, 2-Dihydroquinoline. (TMDQ). The
advantage of the GPC approach becomes evident when
examining the chromatogram of TMDQ oligmers (Figure 4).
This material is a common antioxidant made from the
reaction of aniline and acetone and then polymerized.
If the observed peaks are assigned the normal sequence
- dimer, trimer, tetramer etc., the plot of the retention volume versus the logarithm of the molecular
weights does not produce a smooth line. The peaks
representing the normal loigomeric sequence can be
selected by trial and error and then a different series
of peaks is discovered where the oligomerization
follows a different route. The characterization of
this second series of peaks has been achieved by mass
spectroscopy and reported by Lattimer et. al. (14).

Butylated p-Cresol-Dicyclopentadiene Product. A complex non-staining oligomeric antioxidant marketed by
Goodyear as "Wingstay L" is the butylated reaction
product of p-cresol and dicyclopentadiene. After
trying various analytical techniques, we found that GPC
provides the best approach for following the reaction

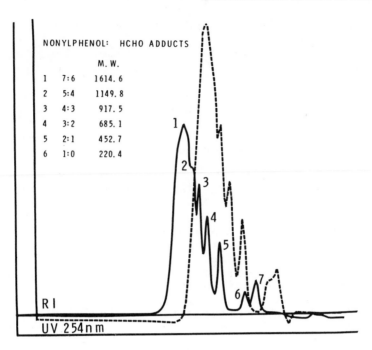

Figure 3. Gel permeation chromatogram of nonylphenol-formaldehyde adducts with dual detectors.

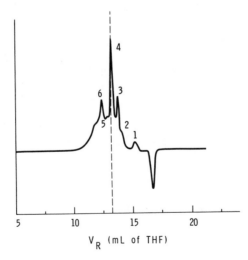

Figure 4. Gel permeation chromatogram of polymerized acetone-aniline condensation products. Key: 3,5- and 6-TMDQ oligomers; and 1,2- and 4-oligomeric series from a different addition route.

products. Both the unbutylated and butylated oligomers
are well separated (Figure 5) and they all fall on a
simple straight line when their retention volumes are
plotted against the logarithm of their molecular
weights. In this case both the differential refractive
index and the UV 254 nm detectors provide essentially
the same information.

Plasticizers. Some of the common plasticizers used for
PVC applications consist of esters of phthalic acid,
epoxidized soya oils and esters of dibasic alkyl acids.
While gas chromatography can resolve some of the lower
boiling materials, a complete analysis can be obtained
by GPC using two detectors. When a complete characterization of the mixture of plasticizers is required, the
use of a UV 254 nm and a differential refractive index
detector is essential. While all the plasticizers
including epoxidized oils respond in the RI detector,
detection of esters with phthalate aromatic moieties
is achieved with the UV 254 nm detector. In a recent
problem where plasticizers were extracted from a polymer were being examined by GPC, we observed a peak
corresponding to di-2-ethylhexyl phthalate. There was
nothing unusual and the identification was confirmed by
GC. A careful examination of the UV and RI responses,
however, showed that the UV response was lower than
expected. The UV/RI ratio for di-2-ethyl-hexyl-phtha-
late is about 2.0 but the sample peak gave a ratio of
only about 1.0.

This information suggested the presence of an
additional component co-eluting with di-2-ethylhexyl-
phthalate. With the information at hand it was
possible to surmize that this component had the following characteristics:

1. It had only minimal response in UV.

2. It had a molecular weight of about 400.

3. It was too high boiling for gas chromatography.

It was then possible to screen some possible can-
didates and come up with the probable presence of a
chlorinated paraffin. This material was found to elute
at the same retention volume as di-2-ethylhexyl-phtha-
late and showed no response at 254 nm (Figure 6). For
quantitation UV/RI response ratio provided all the
data required. However, in order to confirm the presence of this additional component, the sample was
hydrolyzed and the acid converted to its dimethyl ester.
The products were then examined by GPC. The UV response for the peak of interest was completely

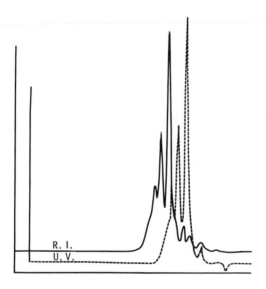

Figure 5. Gel permeation chromatogram of butylated
p-cresol-dicyclopentadiene reaction products.

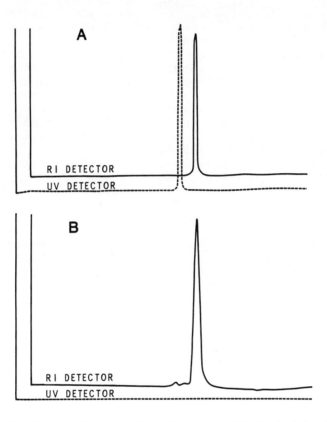

Figure 6.   Gel permeation chromatograms with dual detectors.
Key: A, di-2-ethylhexyl phthalate; and B, chlorinated
paraffin.

eliminated but the RI response due to unhydrolyzable
chlorinated paraffin was still there. The dimethyl
ester of phthalic acid eluted later and showed the
expected UV response.

Quinones and Hydroquinones. In the analysis of quinones and hydroquinones, the use of two different dual
detector systems was required. The retention data for
hydroquinones shows the normal behavior of hydroxyl
groups associating with the solvent, THF. Thus octyl
quinone and hydroquinone elute almost together.
Similarly dioctylquinone and octyl hydroquinone elute
together (Figure 7). The UV/RI response ratio for
benzoquinone is 3.75. Hydroquinone and dioctylquinone
show similar disparities in the UV/RI responses. This
information provides a very good method for detecting
impurities in dioctyl hydroquinone.
    The reverse problem of detecting hydroquinone
impurities in quinones requires the use of an electrochemical detector. The hydroquinones are oxidized at
1.2V to quinones to give very strong responses. The
quinones have no response in this detector.

Poly Chlorinated Biphenyls. The photoconductivity detector provides good responses for polychlorinated
biphenyls separated by GPC. The normal matrix components are detected by RI and UV detectors while the
polychlorinated species show high responses in the
electrochemical detector (Figure 8).

A comprehensive listing of the applications of this
technique in the coatings industry has been presented
in recent publications (15,16).

Conclusions

High-efficiency gel permeation chromatography offers a
very useful technique for the characterization and
analysis of oligomers and small molecules particularly
when multiple detector systems are used.

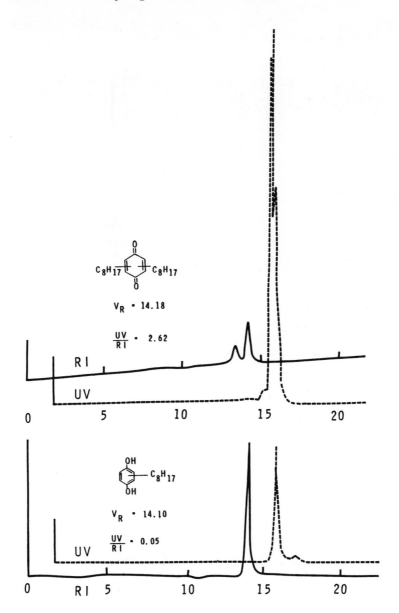

Figure 7.   Gel permeation chromatograms with dual detectors.
Key: top, dioctyl quinone; and bottom, octyl hydroquinone.

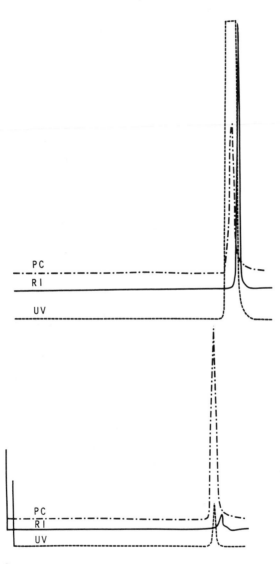

Figure 8. Gel permeation chromatograms with refractive index, UV 254, and photoconductivity detectors. Key: top, p-chlorobiphenyl; and bottom, decachlorobiphenyl.

Acknowledgment

The permission of The Goodyear Tire & Rubber Company to publish this paper is gratefully appreciated.

Literature Cited

1. Hendrickson, J. G., Anal. Chem. 40: 49-53 (1968)
2. Majors, R. E. J., Chromatog. Sci. 18: 488-511 (1980).
3. Hendrickson, J. G. and Moore, J. C., J. Polym. Sci. Part A-1 4:167-88 (1966).
4. Smith, W. B. and Kollmansberger, K., J. Phys. Chem. 69: 4157-61 (1965).
5. Chang, Teh-Liang, Anal. Chim. Acta. 39: 519-21 (1967).
6. Cazes, J. and Gaskill, D. R., Sep. Sci. 2: 421-30 (1967).
7. Cazes, J and Gaskill, D. R., Sep. Sci. 4: 15-24 (1969)
8. Edwards, G. D., and Ng, Q. Y., J. Polym. Sci. Part C. 21: 105-17 (1968).
9. Lambert, A., J. Appl. Chem. 20: 305-306 (1970).
10. Lambert, A, Anal. Chim. Acta. 53: 63-72 (1971).
11. Schulz, W. W., J. Chromatog. 55: 73-81 (1971).
12. Larsen, F. N., App. Polym. Symp. 8: 111-24 (1969)
13. Krishen, A. and Tucker, R. G., Anal. Chem. 49: 898-902 (1977).
14. Lattimer, R. P., Hooser, E. R., and Zakriski, P. M. Rubber Chem. Technol. 53: 346-356 (1980)
15. Kuo, C., and Provder, T. ACS Symposium Series, No. 138: 207-224 (1980).
16. Kuo, C., Provder, T., and Kah, A. F., Paint & Resin, March/April 1983: 26-33.

RECEIVED October 13, 1983

Analysis of Petroleum Crude and Distillates by Gel Permeation Chromatography

C. V. PHILIP and RAYFORD G. ANTHONY

Kinetics, Catalysis and Reaction Engineering Laboratory, Department of Chemical
Engineering, Texas A&M University, College Station, TX 77843

The currently available high efficiency columns
with 5 micron size polystyrene/divinylbenzene
copolymer packing, have extended the capability of
size exclusion chromatography for the separation
of smaller molecular size species in addition to
the large polymeric species. Petroleum crude and
its refinery products are composed of both larger
and smaller molecular components (asphaltenes and
distillates). Gel permeation chromatography (GPC)
using a 100A PL Gel column and tetrahydrofuran
(THF) separates petroleum crude or the refinery
product into fractions containing different chemical species such as nonvolatiles (asphaltenes),
long chain alkanes and aromatics. GPCs of petroleum crude as well as its distillation cuts are
used to illustrate the use of GPC for the analysis
of petroleum crude and its refinery products.

Currently most refineries are capable of processing different
petroleum crudes and can increase the yield of the selected
products on demand. The composition of the crudes varies
depending on factors such as geographical origin, the well
location and depth. Certain ASTM specifications such as API
gravity, viscosity, distillation temperatures, and flash point
are generally used for the evaluations of crude as well as its
refinery products. A number of studies have reported certain
physical and chemical properties of refinery products ([1-5]). It
appears that the ASTM specifications, some of them are a few
decades old, are both time consuming to obtain as well as not
adequate enough to guide the crude through the refinery process
to obtain optimum production of desired distillation cuts.
Thus, other analytical tools to characterize both the crude and
its refinery products are needed.

0097–6156/84/0245–0257$06.00/0
© 1984 American Chemical Society

Gel permeation chromatography (GPC) has been extensively used for molecular size determinations of large molecular weight species such as polymers (6), coal liquids (7-14), and petroleum asphaltenes (15,16,17). GPC data on a number of compounds such as straight-chain alkanes, amines, alcohols, multi-ring aromatics, etc. (7) show that the retention volume is mainly a function of the length of the molecule rather than molecular volume, molecular weight or any other molecular size parameter. The steady increase in the retention volumes of large straight-chain alkanes suggests that they exist in the solution in a stretched state rather than in a coiled state. It is appropriate to say that GPC separations are mostly on the basis of linear molecular size rather than any other molecular size parameter. Longer molecules elute faster than shorter molecules because longer molecules are less likely to diffuse into the liquid trapped inside the pores. The retention volume V_t in a GPC column is given by the equation:

$$V_t = V_i + KV_p \qquad (1)$$

where: V_i = the column interstitial volume; V_p = the total pore volume; and K = the partition coefficient, the ratio of the accessible pore volume to the total pore volume. All solutes elute between V_i and $V_i + V_p$. For most gel columns the value of the ratio of V_i to V_p is in the order of 1 to 1.3. Consequently the total number of peaks that can be resolved by GPC is limited compared to other modes of liquid chromatography. Some molecules are too large for pores of the column and they are eluted without separation at V_i (total size exclusion) and some other molecules are too small and they elute at $V_i + V_p$ (total permeation). By selecting the correct pore size for the molecular size distribution of species in the sample, the resolution of peaks can be increased. Compared to other modes of liquid chromatography relatively larger samples can be separated, without significant loss in resolution. The new columns packed with 5 micron particles have increased the theoretical plate counts (manufacturers claims 40,000 plates/meter) significantly and hence analysis can be accomplished in 10-25 minutes depending on column size and flow rates.

Linear Molecular Sizes from Valence-Bond Structures. In the absence of any interactions between solute and solvent such as hydrogen bonding between solute and solvent molecules resulting in a larger molecular size, and any interactions between solute and gel particles such as adsorption which is the basis for liquid chromatography, the molecular length can be obtained from the valence bond structures. Figure 1 illustrates the fact that rigid molecules such as aromatics are expected to have smaller linear molecular sizes, and consequently, larger retention volumes than straight-chain hydrocarbons of similar molecular weight.

"Effective" Linear Molecular Size in Solutions. When THF is used as the mobile liquid phase, certain species can form hydrogen bonds with THF effectively producing a complex molecule which exhibits a greater linear molecular size (Figure 2) and lower retention volume. When non-polar solvents such as toluene are used, the molecular size is essentially unaffected. Phenol forms hydrogen bonds with THF (Figure 3) resulting in a 1:1 complex and an increase in effective linear molecular size. GPC is widely used for the size separation as well as for the molecular weight distribution of typical polymers. Since molecular length is the main basis for the GPC separation and the fact that the solute size can increase in certain solvents, GPC achieves class separation of species which normally have similar molecular sizes, in some complex mixtures (8,14). Use of GPC for separation of coal liquids into fractions enriched with distinct class of chemical species such as aromatics, phenols, asphaltenes mixed with alkanes, is discussed elsewhere (14).

Experimental

Samples of crude oil and refinery products used in this study were obtained from commericial as well as from local sources. The GPC separations were performed on a Waters Associates Model ALC/GPC 202 liquid chromatograph equipped with a refractometer (Model R401). A Valco valve injector was used to load about 50 microliter samples into the column. A 5 micron size 100A PL gel column (7.5 mm ID, 600 mm long) was used in this study. Reagent grade tetrahydrofuran (THF) which was refluxed and distilled with sodium wire in a nitrogen atmosphere, was used as the GPC carrier solvent. Flow rate was 1 ml per minute. THF was stored under dry nitrogen, and all separations were conducted in a nitrogen atmosphere to prevent the formation of peroxides.
    Straight chain alkanes from Applied Science, aromatics from Fisher Scientific Company and polystyrene standards from Waters Associates were used without purification for the linear molecular size calibration of the GPC. Since the solubility of the larger alkanes in THF is very low, appproximately 0.2 –1 mg of each standard was dissolved in 50 microliters of the THF for the molecular size calibrations.
    The samples of crude and distillate for GPC analysis were prepared by dissolving the sample in dry additive-free THF to obtain a 25% solution and the solution was filtered through micro-pore filters (Millipore, 0.5 micrometer size). A solution containing both the calibration standard and the sample was used to determine the molecular size distribution of the sample.

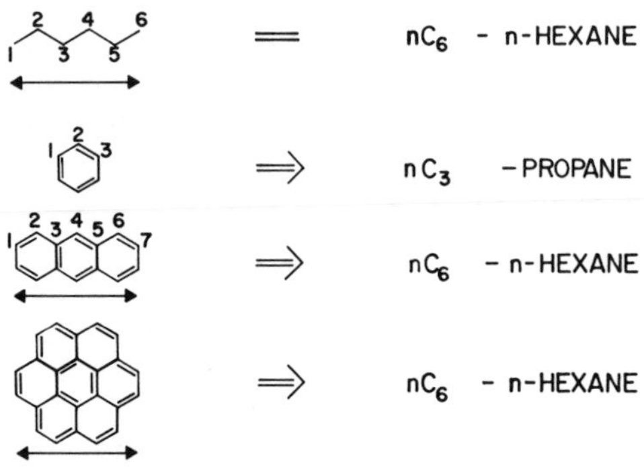

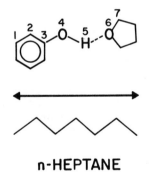

nC_6 – n-HEXANE

nC_3 – PROPANE

nC_6 – n-HEXANE

nC_6 – n-HEXANE

Figure 1. Linear molecules size in straight-chain alkane carbon units with comparable aromatic structures (Fuel, 1982) (14).

n-HEPTANE

Figure 2. Phenol-THF Complex (Fuel 1982) (14).

Results and Discussions

The separation of coal liquids by gel permeation chromatography
using 100A Styragel columns and solvents such as THF and toluene
has been reported elsewhere (7,8,9,13,14). Coal liquids and
petroleum crude are similar in their physical appearance as well
as the complexity in composition. The major difference between
the two is that petroleum crude does not contain oxygenated compounds, such as alkylated phenols, in substantial quantity. In
addition, the average linear molecular size of petroleum derived
asphaltenes (15,16) is much larger than that of coal derived
asphaltenes (9).

Linear Molecular Size: The Best Available Basis for the GPC Separation. The elution pattern of the GPC using 5 micron 100A PL
gel column is illustrated in Figure 4 where the GPC separation
of a standard mixture containing straight chain alkanes and
aromatics is shown. The polystyrene standard (mol. wt. 2350 and
chain length 57A) gave a broad peak at 11 ml retention volume.
The peak position is marked in the figure rather than using the
polystyrene standard in the mixture in order to save the $nC_{44}H_{80}$
peak from the enveloping effect of the broad polystyrene peak.
The rentention volume of several aliphatic and aromatic compounds in THF and toluene have been reported (7). It is clear
that aromatic compounds, as expected from their valence bond
structures, have smaller linear molecular sizes compared to n-
alkanes of similar molecular weight. It is expected that most
of the condensed ring aromatics such as naphthlene, anthracene
and even big ones like coronene (seven fused rings with molecular weight of 300.4) are smaller than n-hexane (14) and hence
have retention volumes larger than that of n-hexane.

The Effect of "Aromatic" Gel on the Size Separation of Aromatic
Species. Certain aromatics such as anthracene, benzopyrene and
coronene produce GPC separation patterns which deviate from what
is expected from their molecular lengths. All aromatic species
have a slightly shorter effective molecular length compared to
their valence bond structures. Although anthracene is about the
size of n-hexane, it has a retention volume close to that of n-
butane. Benzopyrene (five fused rings) has a retention volume
equivalent to propane. The retention volume of coronene (Figure
1) shows that its effective size is slightly smaller than that
of propane. This type of anomalous behavior is expected for a
limited number of compounds due to their structures associated
with extreme aromaticity. The GPC columns are packed with
swelled polymer particles of controlled pore size formed by the
co-polymerization of styrene and divinylbenzene. Every other
carbon atom on the polymer chain has a phenyl group freely
hanging. The species with aromatic structures can interact with
the phenyl groups of the polymer chains of the gel. The inter-

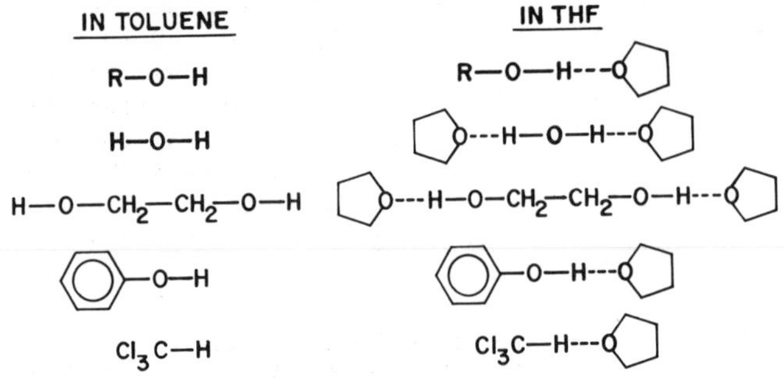

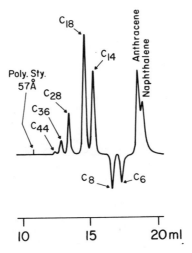

Figure 3. Effect of solvent on the effective linear sizes of molecules in solution (Fuel 1982) (14).

Figure 4. Calibration of GPC retention volume using known compounds.

action between gel particles and the aromatic species can increase
the retention volume slightly which may indicate smaller "effective" molecular length. For samples contining species with non-
uniform structures, the molecular parameter which can be predicted
from the GPC separation pattern is molecular length.

Based on elution volume, the polystyrene standard with a
size of 57A appears to be larger than expected from the alkane
standard. The large number of phenyl groups on the main polymer
chain (57A) make the molecule into a large cylindrical structure
with a large steric hinderance for getting through the pores of
the gel. The two terminal phenyl groups also contribute to an
increase in chain length. The polystyrene peak (57A) is very
close to the total exclusion limit of the 100A PL gel column.
Petroleum crude and its refinery products have two major
component based on distillation. The portion that can be distilled under refinery conditions can be called volatiles and the
nondistillables are the nonvolatiles. The volatiles can be
analyzed by GC or GC-MS. The crude has both components. The
distillate as the names applied, such as naphtha and kerosene
contain only volatiles. When GPC is used for analyzing various
distillates, the fractions separated by GPC can be characterized
by GC or GC-MS. These data can be used to verify the nature of
components present in various distillation cuts as a function of
GPC elution volume. If the samples such as crude contains both
volatiles as well as nonvolatiles, the samples should be
separated into volatiles and nonvolatiles. The GPC of both
components should be used to calibrate the GPC of the total
crude. The parameter that can be obtained from GPC is effective
molecular length. It can be used to relate other molecular
parameters of interest after calibration.
The GPC of a local crude (Bryan, Texas) sample spiked with
a known mixture of n-alkanes and aromatics is shown in Figure 5
and the GPC of the crude is shown in Figure 6. The hydrocarbon
mixture is used to calibrate the length of the species which
separates as a function of retention volume. The molecular
length is expressed as n-alkane carbon units although n-alkanes
represent only a fraction of the hydrocarbons in the crude. In
addition to n-alkanes, petroleum crude is composed of major
classes of hydrocarbons such as branched and cyclic alkanes,
branched and cyclic olefins and various aromatics and nonvolatiles namely asphaltenes. Almost all of the known aromatics
without side chains elute after n-hexane (C_6). If the aromatics
have long side chains, the linear molecular size increases and
the retention volume is reduced. Cyclic alkanes have retention
volumes similar to those of aromatics. GPC separates crude on
the basis of linear molecular size and the species are spread
over 10 to 20 ml retention volume range and almost all of the
species are smaller than the polystyrene standard (57A). In
other words, the crude has very little asphaltenes. The linear

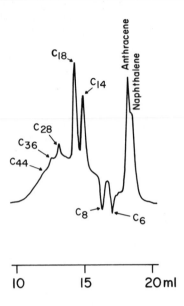

Figure 5. GPC of petroleum crude spiked with known compounds.

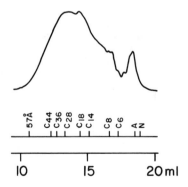

Figure 6. The separation of petroleum crude by GPC.

molecular size distribution of petroleum asphaltene ranges from 50 to 500A with a peak at 100A on the basis of polystyrene standard (15). The bulk of the crude is composed of species with molecular size smaller than that of n-tetratetracontane (nC$_{44}$H$_{90}$). The crude contains a small fraction of aromatics, which is the last peak and has retention volume of 18 ml.

The GPCs of refinery products are shown in Figures 7 to 12. The retention volume, as well as the linear molecular size expressed in n-alkane carbon units are marked in these figures. The refinery products picked for illustrative purpose represent a wide range of distillation cuts as well as nonvolatiles (road asphalt). The commercial grade naphtha (Figure 7) and regular leaded gasoline (Figure 8) have a similar molecular size range, however, regular gasoline has slightly more long molecules. The aromatic regions of the two GPCs may include cyclic alkanes and cyclic olefins in addition to aromatics. Because of the low distillation temperatures of the two, the heavy aromatics are not expected.

GPC of charcoal lighter fluid (Figure 9) shows the increase in the linear molecular size of the cut similar to the ASTM distillation pattern. Lubricating oils are narrower cuts.

The GPC of three oils are shown in Figures 10 and 11. Transmission oil (Dextron II) has a GPC with peaks at n-C$_{16}$H$_{34}$ with a small amount of polymeric additive which separates from the bulk of the oil as a shoulder before the polystyrene mark (57 A). Figure 11 shows the GPCs of two motor oils. Single weight SAE 30 W motor oil has a GPC with a peak at n-C$_{24}$H$_{50}$ with no peaks due to the polymeric species showing the absence of such additives. The multiviscosity motor oil SAE 10-30 W has a GPC peak at n-C$_{17}$H$_{36}$ and a small peak due to added polymeric species. The multiviscosity oil is prepared by adding polymeric additives which make it behave like heavy weight oil at higher temperatures. All these lubricating oils represent narrow distillation cuts from refineries, and it is interesting to note that they have GPCs showing a linear molecular size distribution similar to their distillation temperature distributions or molecular weight distributions. It could be interpreted that these lubricating oils are composed of very similar hydrocarbon species.

The road asphalt used in this study was obtained from the road as a fresh sample. The road asphalt is composed of asphaltenes (GPC peak at 100A and petroleum residual oils (15) (GPC peak at n-C$_{40}$H$_{82}$). The GPC of road asphalt is shown in Figure 9. Since petroleum asphaltenes cannot be separated by a 100A pore size gel column, the asphaltene appears without any separation at the total size exclusion limit of the column. But the nonasphaltene components are separated showing a peak at n-C$_{40}$H$_{82}$. The performance of the road asphalt depends on the asphaltene content as well as on the molecular size distribution of the nonasphaltenic fraction.

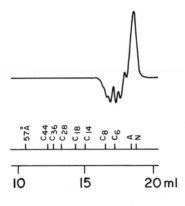

Figure 7.   GPC of naphtha.

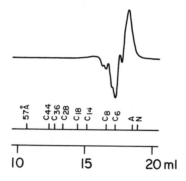

Figure 8.   GPC of regular gasoline.

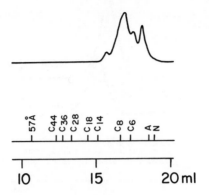

Figure 9.   GPC of charcoal lighter.

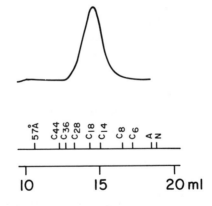

Figure 10.   GPC of transmission oil.

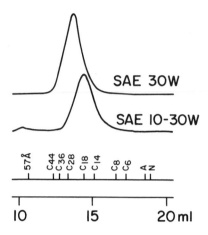

Figure 11.   GPC of motor oils a) SAE 30W, b) SAE 10-30W.

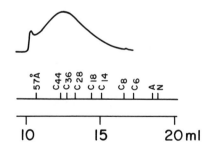

Figure 12.   GPC of road asphalt.

Conclusions

Recently available 5 micron columns such as PL gel columns have increased the efficiency of GPC separations. A single 67 cm colunm packed with 5 micron size 100A PL gel can separate both petroleum crude as well as its refinery products in about 20 minutes consuming about 20 ml of THF. Currently used ASTM specification, especially the most widely used ASTM distillation temperature specifications can be complemented or even substituted by simpler GPC analysis. By comparing the GPC of the crude (Figure 3) with GPCs of various distillation cuts (Figures 4-9) the prediction of the quality of the crude as well as its likely conversion into various cuts could be feasible. The hydrocarbon species eluting at various retention volumes can be identified by the established techniques such as Fourier Transform Nuclear Magnetic Resonance Spectroscopy (FT NMR). The GPC determines the molecular length; FT NMR can continuously monitor effluents for its chemical nature. By using GPC with the appropriate combination of detectors, a very precise evaluation of petroleum crude and predictions of product yield, as well as suitable processing conditions appears to be feasible.

Acknowledgments

The financial support of the Texas Engineering Experiment Station and the Texas A&M University Center for Energy and Mineral Resources is very much appreciated. The authors are grateful to Dr. Jerry A. Bullin for helpful suggestions.

Literature Cited

1. Dice, M. J. "Approximate Calorific Value of California Fuel Oil," Chem. Metal Eng. 1926, 8, 499.
2. Geniesse, J. C. "A Comparison of Viscosity-Index Proposals," A.S.T.M. Bulletin 1956, 7, 81-84.
3. Kryiacopoulos, G. B. "Viscosity Index of Lube Oil Blends," Scien. Lubrication 1962, 5, 27.
4. Kryiacopoulos, G. B. "Blending Oil for Viscosity Index," Hydrocarbon Proc. 1975, 9, 137-138.
5. Woodle, R. A. "Figures Solvent Extract of Heavy Oils," Hydrocarbon Proc. 1966, 45, 133-136.
6. Snyder, R. L.; Kirkland, J. J. "Introduction to Modern Liquid Chromatography"; John Wiley and Sons, Inc., New York, 1974.
7. Philip, C. V.; Anthony, R. G. Am. Chem. Soc. Div. Fuel Chem. Preprints 1979, 24, (3), 204.
8. Philip, C. V.; Zingaro, R. A.; Anthony, R. G. Am. Chem. Soc. Fuel Chem. Preprints 1980, 25, (1), 47.

9. Philip, C. V.; Zingaro, R. A.; Anthony, R. G. in "Upgrading
    of Coal Liquids"; Sullivan, R. F. , Ed.; ACS SYMPOSIUM
    SERIES No. 156, American Chemical Society: Washington,
    D.C., 1981; p. 239.
10. Farnum, S. A.; Olson, E. S.; Farnum, B. W.; Willson, W. G.
    Am. Chem. Soc. Div. Fuel Chem. Preprints 1980, 25 245.
11. Philip, C. V.; Anthony, R. G. Am. Chem. Soc. Div. Fuel
    Chem. Preprints 1977, 22, (5), 31.
12. Philip, C. V.; Anthony, R. G. in "Organic Chemistry of
    Coal"; Larsen, J. W., Ed.; American Chemical Society:
    Washington, D.C., 1978; p. 258.
13. Philip, C. V.; Anthony, R. G. Fuel 1982, 61, 351.
14. Philip, C. V.; Anthony, R. G. Fuel 1982, 61, 357.
15. Philip, C. V.; Bullin, J. A.; Anthony, R. G. "Separation of
    Heavy Fuel Oils by Gel Permeation Chromatography"; sub-
    mitted for publication.
16. Long, R. B. in "Chemistry of Asphaltenes"; Bunger, J. W.;
    Li, N. C., Eds.; ADVANCES IN CHEMISTRY SERIES No. 195,
    American Chemical Society: Washington, D.C., 1981; p. 17.
17. Hall, G.; Herron, S. P. "Chemistry of Asphaltenes"; Bunger,
    J. W.; Li, N. C., Eds.; ADVANCES IN CHEMISTRY SERIES No.
    195, American Chemical Society: Washington, D.C., 1982; p.
    137.

RECEIVED October 13, 1983

COPOLYMERS, BRANCHING, AND
CROSS-LINKED NETWORK ANALYSIS

High-Temperature Size Exclusion Chromatography of Polyethylene

V. GRINSHPUN, K. F. O'DRISCOLL, and A. RUDIN

Guelph–Waterloo Centre for Graduate Work in Chemistry, University of Waterloo, Waterloo, Ontario, Canada, N2L 3G1

The use of dilute polymer solutions for molecular weight measurements requires the macromolecules to be in a true solution, i.e., dispersed on a molecular level. This state may not be realized in certain instances because stable, multimolecular aggregates may persist under the conditions of "solution" preparation. In such cases, a dynamic equilibrium between clustered and isolated polymer molecules is not expected to be approached and the concentration and size of aggregates are little affected by the overall solute concentration. A pronounced effect of the thermal history of the solution is often noted under such conditions.

Stable aggregates have been shown to present a problem in the characterization of polyvinyl chloride (1,2) and it has been suggested that residues of crystalline structures may persist in polyethylene solutions at temperatures below the polymer's crystalline melting point (3-5).

This paper shows the need and describes a method for eliminating aggregates in solutions of polyethylene. In doing so we have developed a simple technique for establishing when polymer solutions are molecularly dispersed. This research is directed toward Size Exclusion Chromatography (SEC) measurements, but low angle laser light scattering (LALLS) has been the primary tool for assessing the quality of the solutions studied.

Experimental

A Chromatix KMX-16 laser differential refractometer and a KMX-6 LALLS photmeter attached to a Waters 150C high temperature gel permeation chromatograph were employed in this work. However, since we were only interested in establishing solution quality, no columns were used in the 150C; only its pump, and automatic injector being employed. Samples of polyethylene solutions, prepared as described below, were injected into the SEC unit at 145°C and their light scattering properties measured as they flowed through the KMX-6, also at 145°C. The KMX-6 was used at

0097–6156/84/0245–0273$06.00/0

an angle of 6--7° with a 0.15 mm field stop to minimize back-
ground. The lasers in both the Chromatix light scattering
photometer and differential refractometer operate at 632.8 nm.
The KMX-6 has a heated flow cell and a heated, insulated tube
through which the sample flows from the liquid chromatograph.

The samples investigated initially were commercial high
pressure low density, linear low density and high density
polyethylenes and had properties given in Table 1. Solutions
of these polymers were prepared in concentrations of 0.8 to 3.5
g/l by dissolving the polymer over a time period of two hours
in an oven maintained at 145°C. To avoid degradation 0.05% 4,4'
-thiobis(3-methyl-6-tert-butyl phenol) was used as an antioxidant in the solutions.

A second set of solutions were prepared from aliquots of the
first solutions by subjecting them to a further thermal treatment of 160°C for 1 hour in an oven.

All solutions were filtered through a 0.5 μm Fluoropore
poly(tetrafluorethylene) filter (FHUP, Millipore Corp.). The
solvents trichlorobenzene (TCB), o-dichlorobenzene (ODCB) and
α-chloronapthalene (αCN) were used as received without further
purification.

Refractive index increments (dn/dc) were measured using the
Chromatix KMX-16 differential refractometer and its heated cell
at 145°C; the values obtained are given in Table II. Refractive index increments are essentially identical for all the
polyethylene types in a given solvent.

Results and Discussion

Figure 1A shows a typical observation of the LALLS for a polymer
solution prepared at 145°C. The many spikes are evidence of
aggregates which have survived the "solution" preparation and
filtering. Heating this solution to 160°C for 1 hour removed
amost all traces of large scatterers (Fig. 1B). This is cir-
cumstantial evidence for the elimination of supermolecular
aggregates. The procedure described below shows quantitatively
that such entities have indeed been removed. It also provides
a criterion for determining if a particular solution history
has provided aggregate-free mixtures.

This is made possible by considering the second virial
coefficients of polymers in the various solvents after
different solution histories. The second virial coefficient,
A_2, can be determined experimentally from the expression

$$\frac{Kc}{R_\theta} = \frac{1}{M_w} + 2A_2c \qquad (1)$$

where c is the solution concentration, K is the polymer
optical constant and is a function of dn/dc, and R_θ is the

Table I. Polyethylene Sample Properties

Designation	Type	Melt Index (g/10 min)	Density (g/ml)
A	High Density	0.25	0.951
B	High Pressure Low Density	0.80	0.921
C	Linear Low Density	1.0	0.920

Table II. Refractive Index Increments (ml/g)

Polymer	Solvent		
	TCB	ODCB	αCN
A	-.112	-.056	-.189
B	-.108	-.051	-.185
C	-.106	-.048	-.180

excess Rayleigh scattering of the solution over the solvent as
measured by the maximum of the ordinate in Figure 1, not considering the pikes, and $\bar{M}_w$ is the weight average molecular
weight.

If the average molecular weight of the sample is known its
second virial coefficient can be predicted using the Kok-Rudin
method (6). Input parameters for this calculation are $\bar{M}_w$,

obtained from LALLS data according to Equation 1, and the Mark-
Houwink constants for the polymer in the particular solvent and
under Theta conditions. The Mark-Houwink constants used in
these calculations are listed in Table III.

The existence of aggregates is evidenced by virial coefficients which are lower than the theoretical values, for the
measured $\bar{M}_w$. This is because the second virial coefficient
decreases with increasing molecular weight. Supermolecular
aggregates appear to have very high effective molecular weights.
$\bar{M}_w$ is relatively little affected by such aggregates in the
concentrations at which they seem to be present. Higher
averages are changed, however, and so is the light scattering
second virial coefficient of the solution.

Experimental and theoretical values of A_2 are compared in
Table IV. The predictions of A_2 and the experimental observations are in good agreement for those experiments in TCB and
ODCB where the sample received the $160^{\circ}C$ treatment and the
"spikes" disappeared.

It was not possible to remove the "spikes" in αCN
solutions by $160^{\circ}C$ treatment. Discoloration occurred if higher
temperatures were used with this solvent. We conclude that αCN
is too poor a solvent for polyethylene to be amenable to this
solution preparation method. In support of this inference, it
may be noted that $\bar{M}_w$ measured in a αCN is always consistently
lower than in the other two solvents. This is presumably because
higher molecular weight molecules are aggregated and appear in
the light scattering trace as "spikes" which do not contribute
to the exceess Rayleigh scattering used to measure $\bar{M}_w$. It will
also be noticed that the second virial coefficient of αCN
solutions remains much lower than in the other two solvents, even
after the prescribed thermal treatment.

Table V compares $\bar{M}_n$, $\bar{M}_w$ and $\bar{M}_z$ values for two polyethylenes
analyzed by SEC in TCB solution at $145^{\circ}C$. Sample C is a linear
low density material listed in Table 1. NBS 1476 is low density
polyethylene which is stated to be a low conversion tubular
reactor product with density 0.931 gcm^{-3} and melt index 1.2 (11).
$\bar{M}_n$ and $\bar{M}_w$ are little affected by the existence of aggregates in
these two samples but $\bar{M}_z$ values are more severely influenced.
It can also be expected that any calculations of long chain
branching frequency (12,13) will be severely compromised by
errors resulting from supermolecular structures. This is because
the frequency of long branches resulting from chain transfer to

Table III. Parameters for Calculation of Second Virial Coefficients*

Solvent	K (cm^3/g)	a		Ref.
TCB	5.96×10^{-2}	0.70		(7)
ODCB	5.06×10^{-2}	0.70	(LDPE)	(8)
ODCB	5.05×10^{-2}	0.693	(HDPE)	(9)
αCN	4.3×10^{-2}	0.67		(9)
Theta	315×10^{-3}	0.5		(10)

*Calculation method is given in (6).

Table IV. Molecular Weights and Virial Coefficients as a Function of Thermal History

Polymer	Solvent	at 145^OC		at 145° with 160° treatment		
		M_w	A_2	M_w	A_2	A_2(theor.)
C	TCB	220,000	4.2×10^{-4}	217,400	1.50×10^{-3}	1.4×10^{-3}
	ODCB	216,100	3.1×10^{-4}	213,700	1.05×10^{-3}	1.09×10^{-3}
	αCN	178,700	3.7×10^{-4}	168,900	7.9×10^{-4}	3.2×10^{-4}
B	TCB	212,300	5.1×10^{-4}	208,300	3.05×10^{-3}	1.51×10^{-3}
	OCDB	215,100	4.2×10^{-4}	210,200	1.71×10^{-3}	1.11×10^{-3}
	αCN	174,200	3.9×10^{-4}	167,300	8.7×10^{-4}	3.4×10^{-4}
A	TDB	–	–	233,600	2.00×10^{-3}	1.41×10^{-3}
	ODCB	–	–	230,800	1.41×10^{-3}	1.05×10^{-3}
	CN	–	–	189,100	6.8×10^{-4}	2.9×10^{-4}

Table V. Comparison of SEC Measurements*

Polymer	145^OC solution			145^OC solution after 160° treatment		
	M_n	M_w	M_z	M_n	M_w	M_z
C(LLDPE)	48,900	188,000	546,500	46,500	185,000	606,800
NBS 1476	27,900	92,400	3,388,000	28,400	93,100	3,722,000

*SEC measurements with 500 Å, 10^4 Å and 10^5 Å Ultrastyragel columns; polymer concentrations were 3.5–5.5 mg/mL in samples injected into TCB at 0.5 mL/min flow rate.

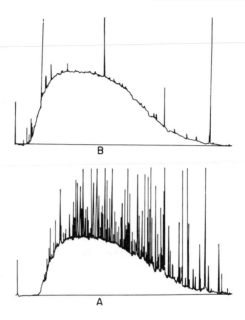

Figure 1.   Strip chart recording of LALLS output for sample
            (A) without and (B) with 160°C treatment. Ordinate
            is scattering an arbitrary units. Sample is
            polyethylene A in TCB at 2.2 g/l and a flow rate
            of 0.1 ml/min with an injection volume of 0.5 ml
            at 145°C.

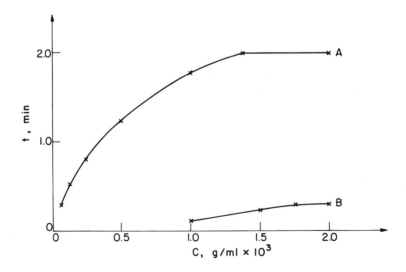

Figure 2.   Time required for maximum scattering to return to
            solvent scattering baseline as a function of polymer
            concentration and flow rate. A flow rate 0.1 ml/min.
            B flow rate 0.5 ml/min.

polymer will probably be greater for higher molecular weight polyethylenes in the high pressure free radical production of this polymer

It is of interest to note in Figure 1 that there is a lag in the response of the LALLS with time caused by mixing in the cell. This is an artifact of the relatively higher viscosity and density of the solution flowing into the cell and displacing the less viscous solvent in the cell. This effect can be seen readily in this work because the solution concentrations were deliberately made high and the flow rate slow. Figure 2 shows the effect of flow rate and concentration on the time required for the cell to be completely emptied of polymer. In conventional SEC measurements this artifact could be of importance, but should not be observed unless very slow flow rates are used. A recent observation on this appears in the work of Rand and Mukherji (14) who used slow flow rates to observe degradation of polymer in an SEC column.

Conclusions

A thermal treatment at $160^{\circ}C$ for 1 hour has proved to be adequate for the removal of aggregates that persist at $145^{\circ}C$ in TCB or ODCB solutions of polyethylene. Such a treatment enables one to obtain true solutions for use in SEC. Solution of polyethylene in αCN appears to be incomplete even after the $160^{\circ}C$ treatment and αCN is therefore not recommended for use in SEC with polyethylene, despite the favorable specific refractive index increment of its solutions.

The 1 hour treatment at $160^{\circ}C$ will be more severe than necessary for some samples and inadaquate for others. We have observed that storage at $160^{\circ}C$ for as much as a day may be required to remove detectable aggregates in solutions of very high molecular weight linear polyethylene samples. In any event, the appropriate duration of such treatments can be assessed by the method discribed here. That is to say, the solution history should be adjusted so that direct measurements of $\overline{M}_w$ by LALLS (without the SEC columns) yields clean recorder traces (as in Figure 1) and second virial coefficients which are in accord with Kok-Rudin predictions of such values for the measured $\overline{M}_w$.

Acknowledgment

Support of this research by the Natural Sciences and Engineering Research Council of Canada is appreciated.

Literature Cited

1. Doty, P, Wagner, H and Singer, S, J. Phys. Chem, 1947, 51, 32.

2.   Rudin, A and Benschop-Hendrychova, I, J. Appl. Polym. Sci.,
     1971, 15, 2881.
3.   Trementozzi, Q.A., J. Polym. Sci., 1959, 36, 113.
4.   Schreiber, H.P. and Waldmann, M.H., J. Polym. Sci., Pt. A,
     1964, 2, 1655.
5.   Stejskal, J. Horska, J. and Kratochvil, P., J. Appl. Polymer
     Sci., 1982, 27, 3929.
6.   Kok, C.M. and Rudin, A, J. Appl. Polym. Sci., 1981, 26,
     3583.
7.   Ram, A and Miltz, J., Appl. Polym. Sci., 1971, 15, 2639.
8.   Dawkins, J.V. and Maddock, J.W., Eur. Polym. J., 1971, 7,
     1537.
9.   Kotera, A., Saito, T., Takamisara, K. and Kiyasara, Y., Rept.
     Progr. Polym Phys. Japan, 1960, 3, 58.
10. Chiang, R., J. Phys. Chem. 1966, 70, 2348.
11. Wild, L., Ranganath, R. and Barlow, A., J. Appl. Polym. Sci.,
     1977, 21, 3331.
12. Foster, G.N., MacRury, T.B. and Hamielec, A.E., in "Liquid
     Chromatography of Polymers and Related Materials II", Cazes,
     J. and Delamare, X., eds., Marcel Dekker, New York, 1980.
13. Axelson, D.E. and Knapp, W.C., J. Appl. Polym. Sci., 1980,
     25, 119.
14. Rand, W.G. and Mukherji, A.K., J. Polym. Sci. Polym. Letters
     Ed., 1982, 20 501.

RECEIVED September 22, 1983

Development of a Continuous Gel Permeation Chromatography Viscosity Detector
For the Characterization of Absolute Molecular Weight Distribution of Polymers

F. B. MALIHI, C. KUO, M. E. KOEHLER, T. PROVDER, and A. F. KAH

Glidden Coatings and Resins, Division of SCM Corporation, Strongsville, OH 44136

A continuous capillary viscosity detector has been developed for use in High Performance Gel Permeation Chromatography (HPGPC). This detector has been used in conjunction with a concentration detector (DRI) to provide information on the absolute molecular weight, Mark-Houwink parameters and bulk intrinsic viscosity of polymers down to a molecular weight of about 4000. The detector was tested and used with a Waters Associates Model 150 C ALC/GPC. The combined GPC/Viscometer instrumentation was automated by means of a micro/mini-computer system which permits data acquisition/reduction for each analysis.

This work describes the design, operation and application of the continuous GPC viscosity detector for the characterization of the molecular weight distribution of polymers. Details of the design and factors affecting the precision and accuracy of results are discussed along with selected examples of polymers with narrow and broad molecular weight distribution.

Recent developments in gel permeation chromatography (GPC) have focused on three major areas including the introduction of high performance columns, instrument automation and the development of molecular weight sensitive detectors. The last area has resulted in the development of laser light scattering photometers,[1,2] and continuous viscosity detectors.[3-5] These detectors when combined with a concentration detector such as a refractive index or an optical density detector in a GPC system, can provide quantitative absolute molecular weight distribution and branching information for polymers. The viscosity detector, although not commercially available, particularly is attractive due to its relative simplicity in design, ease of data reduction and low cost compared to the light scattering detector.

0097–6156/84/0245–0281$06.00/0
© 1984 American Chemical Society

Instrumentation

A schematic diagram of the GPC/Viscometer system is shown in Figure 1. The viscometer is coupled to a Waters Associates Model 150°C ALC/GPC. The key component of the viscometer is a differential pressure transducer (Model P-7D CELESCO, Canoga Park, California) with a ±25 psi pressure range. The transducer monitors the pressure drop across a section of stainless steel capillary tubing (length: 2 ft., I.D.=0.007 in.). The geometric detector volume is about 15 μℓ.

The viscometer assembly is placed in the constant temperature column compartment of the chromatograph between the column outlet and the refractometer. A combination of two Waters Associates M-45 hydraulic filters in series with a capillary tubing coil (length: 10 ft., I.D.:0.01 in.) is used to dampen the line pressure fluctuations caused by the pump. With the above pressure damping modifications the overall system noise was reduced to less than 1 millibar at 1.0 ml/min flow rate in tetrahydrofuran (THF) for a set of six μ-Styragel columns; 10^6, 10^5, 10^4, 10^3, 500, 100Å (Waters Associates, Milford, MA.). The column compartment temperature was set at 50°C.

The automation of the HPGPC/Viscometer system is achieved by interfacing the differential refractometer (DRI) and viscosity detector to a microcomputer for data acquisition. The raw data subsequently, are transferred to a minicomputer (DEC PDP-11/44) for storage and data analysis. Details of the instrument automation are given elsewhere.[6]

Materials

The column set was calibrated with a series of polystyrene standards with weight average molecular weights ($\overline{M}w$) between $2X10^3$ and $4.1X10^6$. The standards were supplied by Pressure Chemical Co., Pittsburgh, Pa. and ArRo Laboratories, Inc., Joliet, Ill. Other systems used in this work included the NBS-706 polystyrene standard and an emulsion polymerized polymethyl methacrylate sample.

Data Reduction

Details of the data analysis for the GPC/Viscometer system have been reviewed by Ouano.[7] The data reduction scheme is summarized in Figure 2 and briefly will be discussed here. The intrinsic viscosity of the effluent at a given retention volume [η](v) is determined from the DRI and continuous viscosity detector responses according to the following equation

$$[\eta](v) = \frac{1}{C(v)} \ln \left(\frac{\Delta E(v)}{\Delta E_o} \right) c \rightarrow o \tag{1}$$

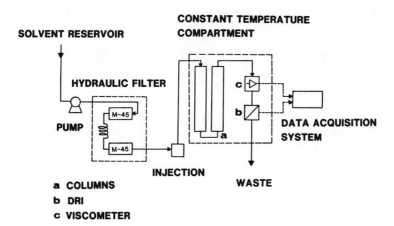

Figure 1.   Schematic of GPC/Viscometer system.

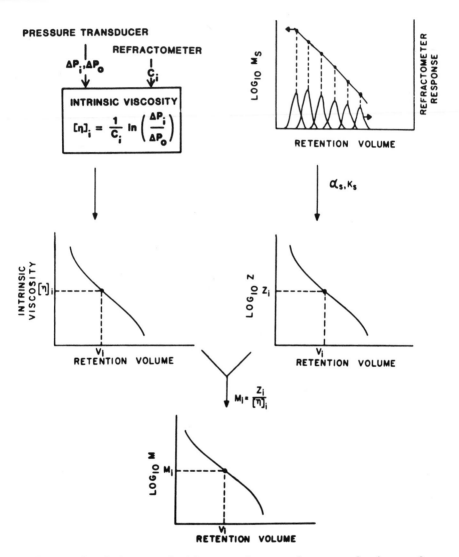

Figure 2. Data reduction scheme for analysis of DRI/Viscometer chromatograms.

where ΔE_0 and $\Delta E(v)$ are the viscosity detector responses (millivolts) at constant flow rate corresponding to solvent and to sample having concentration $C(v)$, respectively. For a linear transducer, $\Delta E(v)$ is proportional to the pressure drop across the capillary, $\Delta P(v)$. The concentration $C(v)$ is given by

$$C(v) = W \cdot f(v) / \int_{v_L}^{v_H} f(v)dv \qquad (2)$$

where W (grams) is the weight of the sample injected, and $f(v)$ is the concentration detector (DRI) response at the retention volume v. The parameters v_L and v_H represent the lowest and highest retention volume in the chromatogram.

From the primary calibration curve based on polystyrene standards and the Mark-Houwink constants for polystyrene (K,α) a universal calibration curve (Z vs. v), based on hydrodynamic volume is constructed. Z is calculated from

$$Z(v) = [\eta](v) \cdot M(v) = KM(v)^{\alpha+1} \qquad (3)$$

Using the intrinsic viscosity data and the universal calibration curve(8) a secondary molecular weight calibration curve can be constructed for the polymer of interest as shown by the following equation:

$$M_x(v) = Z(v)/[\eta]_x(v) \qquad (4)$$

From this information the absolute molecular weight distribution and the intrinsic viscosity-molecular weight plot can be constructed. From this plot the solvent and temperature dependent Mark-Houwink coefficients for linear polymers and information for polymer chain-branching of non-linear polymers can be obtained.

Effect of Operational Parameters

Successful operation of the viscometer depends on good control of possible sources of flow variations in the system which include pump pulsations, temperature variations and restrictions in the GPC columns and fractional sections of tubing.

Pump Pulsations

The noise due to the reciprocating action of the dual-headed pump in the 150C ALC/GPC has a constant frequency pattern and can be reduced by hydraulic filters as shown in Figure 3. A 60% reduction in the peak-to-peak noise is achieved by using a Mark II dampener (Laboratory Data Control, River Beach, Florida), while the M-45 filters under same conditions have reduced the noise by 90%.

Flow Rate

Figure 4 shows the effect of flow rate on the stability of the viscometer baseline signal. Results indicate that the increase in flow rate reduces the high frequency noise (pump noise) while increasing the low frequency noise, apparently caused by imperfections in the flow system (e.g., column packing condition, end fittings, adsorbed sample impurities, etc.). Optimum operating conditions can be established for flow rates between 1 to 1.5 ml/min. Similar results were obtained when DuPont Zorbax Bimodal columns (DuPont Co., Wilmington, DE.) and Varian MicroPak TSK columns (Varian Associates Inc., Palo Alto, Calif.) were used with the GPC/Viscometer system.

Imperfections in the Hydraulic System

The low frequency baseline noise of the viscometer can be substantially reduced by careful filtration of samples and regular checking and maintenance of column end fittings and fractional sections of tubing in the system. Figure 5 shows the effect of column screen replacement on the stability of the baseline signal at a flow rate of 1.0 ml/min.

Sensitivity of the Viscometer

Once the major sources of the viscometer baseline noise are eliminated the viscometer detector can be used for the analysis of polymers with molecular weights as low as 2000. The precision of the GPC/Viscometer analysis is influenced to a great extent by the signal-to-noise ratio (S/N) of the viscometer response at each point on the chromatogram. An estimate of the S/N ratio for this viscometer system is provided in the chromatograms shown in Figures 6 and 7. In these Figures the viscometer response is shown for two narrow MWD standard polystyrene samples with average molecular weights of 4,000 and 97,000, with injected mass of 1.50 mg and 0.690 mg, respectively.

At a flow rate of 1 ml/min with THF as mobile phase at 50°C, initially a baseline pressure of 664 millibars is established for both samples. For the 4,000 molecular weight polystyrene a maximum pressure of 668 millibars is reached at the peak position

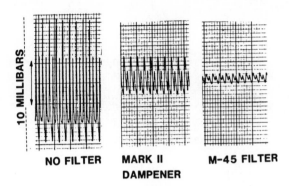

Figure 3.   Effect of the commercial hydraulic filters on the baseline noise of the viscometer trace (Mobile Phase: THF, Flow Rate: 0.5 ml/min, Temperature: 30°C)

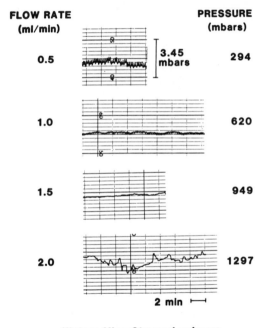

Figure 4.   Effect of flow rate on the baseline noise of the viscometer trace (Mobile Phase: THF, Temperature: 30°C)

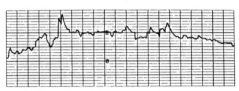

BEFORE SCREEN REPLACEMENT

AFTER SCREEN REPLACEMENT

Figure 5. Viscometer trace before and after the column screen replacement.

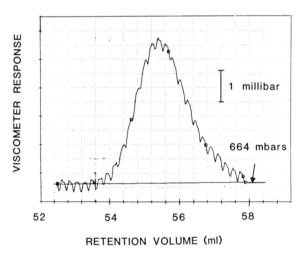

Figure 6. Viscometer chromatogram of narrow MWD standard polystyrene. (Mw:4000, Injected Mass: 1.50 mg, Flow Rate: 1 ml/min, Temperature: 50°C, Column: as in Figure 4)

with 0.32 millibar peak-to-peak baseline noise. Results for the 97,000 molecular weight sample indicate a maximum pressure of 679 millibars, with a peak-to-peak noise of 0.45 millibar. At the point of maximum pressure the S/N for the 4000 molecular weight sample is 10.5 and for the 97000 molecular weight sample is 32 indicating good S/N performance for both systems.

The high sensitivity of the viscosity detector to the high molecular weight fractions is demonstrated in the analysis of a sample of very high molecular weight poly(methyl methacrylate) shown in Figure 8. A shoulder at 3,000,000 molecular weight detected by the DRI becomes a peak when detected by the viscometer detector.

Determination of the Dead Volume Between the Viscometer and the Concentration Detector

Another requirement for accurate GPC/Viscometer data analysis is accounting for the dead volume (ΔV) between the viscometer and the concentration detector.

Reported literature(10) and our own experience have shown that an estimate of ΔV based on the geometry of the connecting tubing is not reliable for this purpose. This primarily is due to the variation in the internal diameter of the commercially available tubing. In this work we have applied a semiempirical experimental method to determine ΔV. The method recently has been implemented by Lesec and coworkers.(10)

In this method one injects a known amount of a high molecular weight polymer on to low porosity GPC columns. From the viscometer and DRI chromatograms, as shown in Figure 2, the apparent intrinsic viscosity $[\eta](v)$ is determined and plotted against retention volume v. A series of $[\eta](v)$ vs. v plots are then constructed assuming a range of dead volumes. The slope of each plot is determined by linear regression and is plotted against the assumed ΔV. The correct ΔV corresponds to the zero slope.

To implement this technique a combination of two μ-styragel columns with 100Å and 500Å porosity was used. A sample size of 50$\mu\ell$ of 0.1% (W/V) standard narrow distribution polystyrene with $\bar{M}_w = 1.8 \times 10^6$ was injected on to the columns. THF was used as the mobile phase at a flow rate of 1 ml/min and temperature of 50°C. Using the data analysis routine described above, a value of 115$\mu\ell$ was obtained for ΔV as shown in Figure 9.

Quantitative Analysis

Figure 10 shows DRI and viscometer traces for the NBS 706 polystyrene standard. Based on the information from these two chromatograms in conjunction with the universal calibration curve, one can calculate the intrinsic viscosity $[\eta](v)$ and molecular weight M(v) at each retention volume as shown in

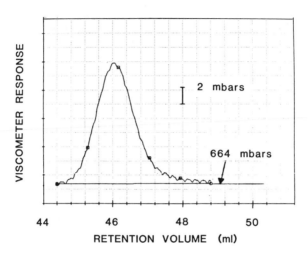

Figure 7. Viscometer chromatogram of narrow MWD standard polystyrene. (Mw:97,000, Injected Mass: 0.690 mg, Other conditions: same as in Figure 6.)

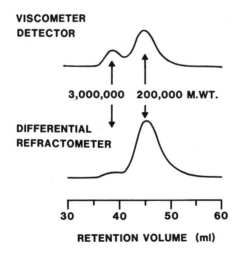

Figure 8. DRI and viscometer chromatograms for a high molecular weight polymethyl methacrylate sample.

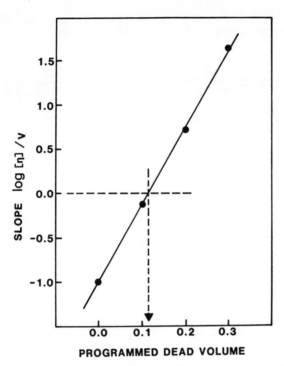

Figure 9. Slope versus ΔV plot for the determination of the dead volume between DRI and viscometer detectors.

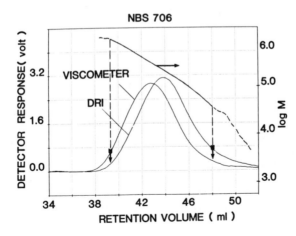

Figure 10. DRI and viscometer chromatograms, and log (M) vs. (v) for NBS-706. (Mobile Phase: THF, Flow Rate: 1.0 ml/min, Temperature: 50°C, Injected Mass: 2.15 mg, Column Set: as in Figure 4.)

Figures 10 and 11. The molecular weight M(v) can then be used to generate molecular weight distribution statistics as summarized in Table I.

Table I.   Molecular Weight Distribution Statistics For NBS-706

	$\bar{M}_n$	$\bar{M}_w$	$\bar{M}_w/\bar{M}_n$
From Secondary Calibration	133,900	278,400	2.08
From NBS Data Sheet	136,500	257,800	1.89

Figure 12 shows the classical method of obtaining the Mark-Houwink coefficients, K and α, by plotting the log [η](v) vs. log M(v) for this polymer in THF at 50°C. The data points used for the plot in Figure 12 are indicated by the area between the arrows in Figure 10. Linear regression analysis of the data resulted in $K_{50^\circ C}=1.86 \times 10^{-4}$ and $\alpha_{50^\circ C}=0.662$ with a correlation coefficient of R=0.9996 for NBS 706 polystyrene.

Table I indicates good agreement between the molecular weight distribution statistics obtained by coupled GPC/Viscometer method and the nominal values for NBS 706. The discrepancy between the Mark-Houwink parameters obtained here and the reported values for polystyrene standard (9) in THF at 25°C (i.e., α = 0.706 and k = 1.60 x 10^{-4}) may in part be due to the uncertainty involved in the determination of the dead volume between DRI and viscometer detectors. Our simulation studies over a range of dead volume values (0 to 120µℓ) showed that α and k are quite sensitive to the dead volume between the detectors. Larger dead volume results in smaller α and larger k values. This is a direct result of a clockwise rotation of log [η] vs. log M(v) curve (Figure 12) which occurs when the dead volume correction is applied in quantitative analysis. The effect on the molecular weight statistics, however, appeared to be small with $\bar{M}_n$ being more sensitive to this correction.

This simulation study also indicated that the result of the experimental determination of dead volume (as described above) may be overestimated by as much as 50%. A possible reason for this could be due to the lack of complete exclusion of the high molecular weight polystyrene used in this technique. Further refinement of this semi-empirical experimental technique for the determination of the dead volume is needed. Details on the simulation studies will be reported in future communications.

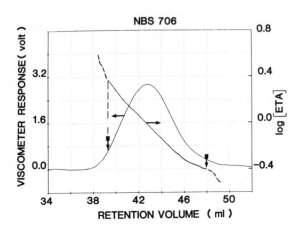

Figure 11. Viscometer chromatogram and log [ETA] vs. (v) for NBS-706. (Conditions as in Figure 10).

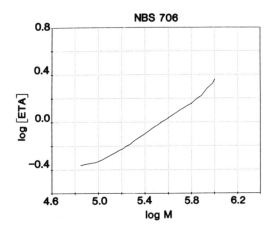

Figure 12. Log [ETA] vs. log M for NBS-706. (Conditions as in Figure 10).

Other possible reasons for the discrepancy in the Mark-Houwink parameters may be due to the band spreading effects and inadequate signal-to-noise quality at the tails of the viscometer chromatogram. These subjects will be the topic of our future investigations in this area.

Summary

The use of a continuous GPC viscosity detector in conjunction with a DRI detector permits the quantitative determination of absolute molecular weight distribution in polymers. Furthermore, from this combination one can obtain Mark-Houwink parameters and the bulk intrinsic viscosity of a given polymer with a GPC calibration curve based only on polystyrene standards. Coupling these two detectors with ultraviolet and infrared detectors then will permit the concurrent determination of polymer composition as a function of molecular weight and branching. This work will be reported in future communications.

Literature Cited

1. A. C. Ouano and W. Kaye, J. Polym. Sci. Polym. Chem. Ed., 12. 1151 (1974).
2. M. L. McConnell, Am. Lab., 10(5), 63 (1978).
3. A. C. Ouano, J. Polym. Sci., A-1, 10, 2169 (1972).
4. L. Letot, J. Lesec, C. Quivoron, J. Liq. Chromatogr. 3, 427 (1980).
5. F. B. Malihi, M. E. Koehler, C. Kuo, and T. Provder, 1982 Pittsburgh Conference, Paper No. 806.
6. M. E. Koehler, A. F. Kah, T. F. Niemann, C. Kuo and T. Provder, "An Automated Data Analysis System for A Waters Model 150C ALC/GPC System with Multiple Detectors", This Volume.
7. A. C. Ouano, J. Macromol. Sci., Revs. Macromol. Chem., C9(1), 123 (1973).
8. T. Provder and E. M. Rosen, Separation Sci., 5 (4), 437 (1970).
9. T. Provder, J. C. Woodbrey, J. H. Clark, E. E. Drott, Advances in Chemistry Series No. 125 "Polymer Molecular Weight Methods", Ed. E. Ezrin 117 (1973).
10. D. Lecacheux and J. Lesec, J. Liq. Chromatogr., 12, 2227, (1982).

RECEIVED October 4, 1983

Size Exclusion Chromatography with Low-Angle Laser Light-Scattering Detection

Application to Linear and Branched Block Copolymers

R. C. JORDAN, S. F. SILVER, R. D. SEHON, and R. J. RIVARD

3M—3M Center, P.O. Box 33221, St. Paul, MN 55133

Linear block polymers of styrene–co–isoprene and
styrene–co–butadiene were prepared via anionic
polymerization and subsequently were coupled
with divinyl benzene to give multi–arm macromolecules. Low–angle laser light scattering
(LALLS) was used for molecular weight measurement both in the stand–alone (static) mode and
as a detector coupled to SEC. No dependence of
the weight average molecular weight ($\overline{M}_W$) on
solvent was found, which is consistent with light
scattering theory for compositionally homogeneous
block polymers. Comparison of SEC/LALLS data
for both the linear and branched species shows
the strong effect of branching upon the hydrodynamic volume/molecular weight relationship.
The data indicate the multi–arm samples are of
relatively small molecular weight polydispersity,
with a weight–average branching functionality of
16–18. Use was made of the universal calibration
and SEC/LALLS data to calculate the branched/
linear intrinsic viscosity ratio through the
molecular weight distribution of the multiarm
samples; an anomalous dependence on molecular
weight was found. Difficulties with the universal
calibration procedure or sample viscosity effects
are discussed as possible causes.

The deliberate introduction of multifunctional branching into
anionically prepared polydiene and poly(diene–co–styrene)
polymers produces materials with unique morphological and
viscoelastic properties (1–3). Work has included synthesis of
symmetric star polymers produced by reaction of living
polyanionic "arms" with multi–functional chlorosilane (4–9),

0097–6156/84/0245–0295$07.50/0
© 1984 American Chemical Society

and less well defined structures have been obtained using divinyl benzene (DVB) as the linking agent (10–16).

Dilute solution studies on such materials have been carried out in an attempt to gain some understanding of molecular structure/property relationships (6, 7, 14, 17–19), and the work has benefited from elegant theoretical frameworks developed by several workers (20–24). Molecular weight characterization is critical to fundamental investigations, as well as to development of controlled production processes in commercial applications.

Several studies have been published which utilize size exclusion chromatography (SEC) for characterization of the molecular weight distribution of multi–arm structures of polystyrene, polyisoprene, and block copolymers of styrene/butadiene and styrene/isoprene (1, 2, 8, 17, 25–26). An interesting phenomenon of direct consequence to work presented here was recorded by Bi and Fetters (1) as a result of their work on DVB–linked "star" block copolymers of poly(styrene–co–polybutadiene) and poly(styrene–coisoprene): they found that the universal calibration procedure gave anomalous molecular weight values for several samples. These workers noted that the sample intrinsic viscosity ([η]) was insensitive to the number of arms, when the arm composition and molecular weight was held constant, and it was suggested that this characteristic might underlie the unusual chromatographic behavior. However, it was noted that some degree of SEC separation by number of arms was achieved. These authors suggest that unique hydrodynamic properties conferred by a "core/shell" structure may be responsible for the SEC behavior, and also it is not clear what relative effects the number of arms and arm length have on the SEC separation.

The observations and questions raised by these workers helped stimulate the experiments which are presented here. In this work, size exclusion chromatography with low angle laser light scattering detection (SEC/LALLS) was used to probe the distribution of branching functionality in multiarm macromolecules of styrene/isoprene and styrene/butadiene block copolymers and to evaluate the applicability of universal calibration to analysis of such materials. The SEC process separates via hydrodynamic volume, not molecular weight, so that use of conventional calibration methods (e.g., calibration with linear polystyrene standards) suffers from the inherent hydrodynamic volume/molecular weight differences between sample and calibrant. In general, branching decreases the hydrodynamic volume of a macromolecule relative to its linear homolog, so that conventional SEC analysis of branched species is especially precarious. Use of LALLS with SEC can circumvent such difficulties since connection of a LALLS detector in series with a concentration detector allows determination of the correct molecular weight at each increment in the polymer chromatogram without recourse to use of calibration techniques (27-29). The SEC/LALLS method has proved to be a valuable tool for the study of branched polymers, with several excellent publications appearing on this subject (17, 26, 30-33).

Synthetic copolymers often present a fundamental difficulty for molecular weight measurement by light scattering, since compositional heterogeneity can be superimposed on the distribution of molecular weights. Whereas a dilute solution light scattering measurement from a homopolymer which is polydisperse in molecular weight yields the weight-average molecular weight of the sample ($\overline{M}_W$), a compositionally heterodisperse polymer sample gives an apparent molecular weight which depends on the solvent (34). The copolymer difficulty arises from the change of the specific refractive index increment (dn/dc) with polymer composition, while a value of dn/dc measured on the entire sample is used to calculate molecular weights from light scattering data. In cases without compositional variation among polymer chains, the light scattering molecular weight is independent of solvent identity.

The experimental samples used in this work should, however, be amenable to straightforward light scattering analysis since the constituent polymers possess backbone block microstructures which guarantee compositional uniformity throughout the sample. The synthetic route utilizes "living" anionic polymerization of polystyrene, followed by isoprene or butadiene addition; divinyl benzene (DVB) linkage of these species gives multiarm structures (MA) containing arms which are homogeneous with respect to composition and molecular weight. The light scattering average molecular weight of both the linear block (LB) copolymer and MA should be independent of solvent identity, with an identical dn/dc for both materials.

The SEC/LALLS method was 1) used to study a commercially available linear block copolymer: Kraton 1107 brand elastomer (Shell Chemical Company), and 2) to study the starting "arm" block copolymer and resultant DVB-linked MA of several experimental samples.

Theory

Dilute Solution Light Scattering: Homopolymers and Copolymers. The theoretical basis for polymer molecular weight measurement by light scattering has been developed in detail, and only the concepts relevant to SEC/LALLS studies of copolymers are presented here.

Proper use of LALLS to measure polymer molecular weight requires a dilute solution of optically isotropic flexible macromolecules whose dimensions are of the same

order as the wavelength of scattering radiation (35-37). For
chains which are compositionally and molecular-weight monodisperse, in the limit of vanishingly small observation angle
(θ):

$$\frac{Kc}{\overline{R}_\theta} = \frac{1}{M} + 2A_2 c \qquad (1)$$

where $\overline{R}_\theta$ is the "excess" Rayleigh factor calculated from the
excess scattering of polymer solution over solvent, and A_2
and c are the second virial coefficient and polymer
concentration, respectively. The quantity K is an optical
constant, defined for the polarized laser light source and
particular annular collection optics of the commercially
available LALLS detectors as:

$$K = \frac{2\pi^2 n_0^2}{\lambda_0^4 N_A} (1 + \cos^2 \theta)(\frac{dn}{dc})^2 = K'(\frac{dn}{dc})^2 \qquad (1a)$$

where n_0, λ_0, and N_A represent the solvent refractive index,
in vacuo scattering wavelength, and Avogadro's number,
respectively (38, 39).

The quantity dn/dc is the specific refractive index
increment and it represents the incremental change in
solution refractive index with sample concentration at the
wavelength, temperature, and pressure of the LALLS measurements. Since dn/dc reflects the optical characteristics of the
polymer and solvent (their different optical polarizabilities),
its value strongly depends on the chemical composition of
both components (40).

In block copolymers containing monomers A and B, to a
good approximation the overall dn/dc is (40):

$$(\frac{dn}{dc}) = W(\frac{dn}{dc})_A + (1-W)(\frac{dn}{dc})_B \qquad (1b)$$

where the subscripted parameters refer to the dn/dc of
homopolymers of A and B, measured in the same solvent and
at the same temperature and wavelength as the copolymer.
The weight composition W is:

$$W = \frac{M_A}{M_A + M_B}$$

where M_A is the molecular weight of the polymer of only the
A subunits of the copolymer and M_B is the corresponding
quantity for B subunits.

The dn/dc can show a dependence on polymer tacticity
(41) and molecular weight, but these effects usually are
minor relative to that of polymer composition (40). Also, the

magnitude of dn/dc increases with decreasing wavelength (42) as $(1/\lambda_o)^2$, and it shows a small dependence on temperature; the quantity must be measured at the same wavelength and temperature as used in the scattering measurement (40).

Next, consider a polymer sample which is heterodisperse both in molecular weight and composition. In the limit of vanishing concentration, Equation 1 gives, for independent scatterers, i:

$$(\overline{R}_\theta)_i = K'c_iM_i(\frac{dn}{dc})_i^2 \tag{2}$$

where from Equation 1a we see that:

$$K' = \frac{2\pi^2 n_o^2}{\lambda_o^4 N_A}$$

which is independent of molecular identity. Assuming that the total excess Rayleigh factor is the sum of individual scatterers, Equation 2 gives:

$$\overline{R}_\theta = K' \sum_i c_iM_i(\frac{dn}{dc})_i^2$$

or

$$\overline{R}_\theta = K'c(\frac{dn}{dc})^2 \left[\frac{\sum c_iM_i(\frac{dn}{dc})_i^2}{c(\frac{dn}{dc})^2}\right] = K'c(\frac{dn}{dc})^2 M* \tag{3}$$

where c and (dn/dc) are the sample concentration and specific refractive index increment, respectively. The quantity M* is an apparent average molecular weight, and it will vary with solvent identity because of the compositional (and associated dn/dc) differences in individual molecular species i.

The light scattering equation for molecular weight-heterodisperse samples which are compositionally homogeneous simplifies if dn/dc is constant for all species:

$$\overline{R}_\theta = K'c(\frac{dn}{dc})^2 \overline{M}_w = Kc\overline{M}_w \tag{4}$$

The derivation of Equation 4 utilizes the definition of the weight-average molecular weight:

$$\overline{M}_w = \frac{\sum c_iM_i}{\sum c_i}$$

Hence the equation for LALLS measurement of the $\overline{M}_w$:

$$\frac{Kc}{R_\theta} = \frac{1}{\overline{M}_w} + 2A_2 c \qquad (5)$$

For copolymers, the above development shows that LALLS molecular weight measurements can be carried out in several solvents in order to check for compositional polydispersity. Polymers which are compositionally homogeneous will give a $\overline{M}_w$ which is independent of solvent identity.

Size Exclusion Chromatography with Low-Angle Laser Light Scattering (SEC/LALLS). A size exclusion chromatograph with both LALLS and concentration detectors gives the correct weight-average molecular weight M_w (v) of polymers with concentration c(v) in elution volume v (28-31, 33). With adequate SEC resolution, M_w (v) represents the molecular weight of a species which is monodisperse in molecular weight M. In all that follows, we assume that the latter condition is approximated and that M_w(v) = M(v). The reasonableness of this assumption will be examined in the context of data for samples analyzed in this work. The fundamental LALLS equation (Equation 5) forms the basis of the SEC/LALLS method.

Universal Calibration. One of the goals of this work was to evaluate the applicability of the universal calibration technique (43) to SEC analysis of these multi-arm macromolecules. This technique assumes a unique calibration relationship

$$[\eta](v) \cdot M(v) = J(v) \qquad (6)$$

for the SEC system which describes the elution behavior of all samples. In Equation 6, $[\eta](v) \cdot M(v)$ is the product of the intrinsic viscosity ($[\eta]$) and molecular weight (M) of a molecular-weight monodisperse polymer eluting in v. The relationship in Equation 6 first was proposed and demonstrated by Benoit et al. (44) and shown to hold for polymers with a spectrum of configurations, including rod-like, branched, and linear random coil structures. However, as noted above, studies of multi-arm stars of linear poly(diene-co-styrene) arms suggest deviation from universal calibration behavior (1).

Branching Parameter g' from SEC/LALLS. The effect of polymer branching upon the dilute solution configuration of polymers is conveniently expressed as the ratio of intrinsic viscosities of branched and linear polymers of the same chemical composition and molecular weight (35), i.e.,

$$g' = (\frac{[\eta]_b}{[\eta]_l})_M \tag{7}$$

where subscripts b and l refer to branched and linear material, respectively, and the subscript M denotes constant molecular weight.

In order to determine g' as a function of molecular weight, one approach is to use universal calibration with SEC analysis of molecular–weight polydisperse samples (31–33). For a multiarm (MA) branched material, the intrinsic viscosity of polymer eluting in v is:

$$[\eta]_{MA}(v) = \frac{J(v)}{M_{MA}(v)} \tag{8}$$

where $M_{MA}(v)$ is the light scattering molecular weight in v. Also, from the Mark–Houwink relationship (43):

$$J(v) = K_{PS}(M_{PS}(v))^{1+a_{PS}} \tag{9}$$

where K_{PS} and a_{PS} denote the Mark–Houwink parameters for polystyrene calibrants in the chromatographic solvent, and $M_{PS}(v)$ is the molecular weight (weight–average) of a narrow distribution standard with peak elution volume v.

Now consider MA materials which consist of linked arms of identical linear block copolymers (LB). Define the ratio:

$$k = \frac{M_{PS}(v)}{M_{LB}(v)} \tag{10}$$

which represents the difference in molecular weight/elution volume behavior for polystyrene calibrants and LB. We assume k is constant over the calibration range. These relationships can be used to calculate the intrinsic viscosity of LB material which has the same molecular weight as MA eluting in v $(M_{MA}(v))$; the LB will elute in some earlier volume v_e. The molecular weight of polystyrene eluting at v_e is k x $M_{MA}(v)$, and the universal calibration relationship gives:

$$[\eta]_{LB}(v_e) = \frac{K_{PS}(k \times M_{MA}(v))^{1+a_{PS}}}{M_{MA}(v)} \tag{11}$$

The viscosity ratio g'(v) then can be defined for MA eluting in v. Using Equations 11 and 10 in Equation 7 gives:

$$g'(v) = \left[\frac{M_{PS}(v)}{k \times M_{MA}(v)} \right]^{1+a}_{M_{MA}(v)} PS \tag{12}$$

The above derivation rests on three assumptions:
1) Validity of the universal calibration,
2) SEC system performance sufficient to resolve the polymer sample into discrete molecular weight species at each v; i.e., band spreading is negligible and chromatographic artifacts such as "viscous streaming" (43) are absent, and
3) a constant value of the polystyrene/LB molecular weight ratio through the chromatogram.

Experimental

Polymer Synthesis/Materials. Multiarm samples were prepared via anionic polymerization in cyclohexane at 50-60 deg C. Polystyrenyl lithium anions of desired molecular weight were prepared with S-butyl lithium initiation, followed by addition either of isoprene or butadiene to give block polydienyl anion. A sampling of the latter was taken, terminated, and used as representative LB arm. The MA samples were synthesized by addition of DVB at a mole ratio of 4.5 DVB:1 anion. The polymerization was terminated by methanol addition. Four styrene/isoprene LB samples (SI-X) of different molecular weight and composition were prepared along with the corresponding MA: (SI-X) DVB; one styrene/ butadiene (SB-1) and its MA ((SB-1) DVB) was made. Proton NMR gave the following weight percent styrene for each LB: SI-1 (9%), SI-2 (23%), SI-3 (26%), SI-4 (48%), and SB-1 (53%).

Homopolymers of butadiene, isoprene, and styrene were prepared under similar conditions. It should be noted that DVB was a commercial grade and, therefore, consisted of meta/para isomers and ethyl vinyl benzene.

Kraton 1107 brand elastomer was from Shell Chemical Co., and it is synthesized by coupling the isoprenyl anion ends of a styrene/isoprene (SI) block copolymer to give styrene/isoprene/styrene (SIIS). Proton NMR analysis indicated 84% (wt.) isoprene and 16% (wt.) styrene.

Polystyrene calibration standards were from Pressure Chemical Co. and all had polydispersities ($\overline{M}_w/\overline{M}_n$) less than 1.1.

SEC System, Data Processing, and Chromatography Procedures. The SEC/LALLS system contained a Model 110A pump (Altex), Model 7125 injector (Rheodyne), KMX-6 Low-Angle Laser Light Scattering Photometer (LDC/Milton Roy), and a Model 98.00

Refractive Index Detector (Knauer). The KMX-6 scattering intensity was measured with the 6-7 degree forward-scattering annulus. A series of Zorbax PSM columns (DuPont) was used: PSM 60, PSM 1000, PSM 1000, PSM 60, PSM 1000. Tetrahydrofuran (THF) from Baker was filtered through a 0.22 micrometer Fluoropore filter (Millipore Corp.) before use in chromatography, and a flow rate of 0.7 ml/min was used.

Analog detector data were acquired via analog/digital Instrument Interface Modules (LDC/Milton Roy) connected in series to a Minc 11/23 (Digital Equipment Corp.) computer. Software packages for run-scheduling and data acquisition ("RTDAS-I"), conventional calibration SEC ("GPC-II"), and SEC/LALLS data processing ("MOLWT-II") were from LDC/Milton Roy.

The MOLWT-II program calculates the molecular weight of species in retention volume $v(M(v))$, where v is one of 256 equivalent volumes defined by a convenient data acquisition time which spans elution of the sample. Moments of the molecular weight distribution (e.g., $\overline{M}_z$, $\overline{M}_w$, $\overline{M}_n$) are calculated from summation across the chromatogram. Along with injected mass and chromatographic data, such as the flow rate and LALLS instruments constants, one needs to supply a value for the optical constant K (Equation la), and second virial coefficient A_2 (Equation 1). The value of K was calculated for each of the samples after determination of the specific refractive index increment (dn/dc) for the sample in the appropriate solvent. Values of A_2 were derived from off-line (static) determinations of $\overline{M}_w$.

A universal calibration curve was developed, using the retention volume v_m corresponding to the DRI detector peak maximum of eluting polystyrene calibrants. Data were fitted with the GPC-II program to an equation of the form:

$$\ln J(v_m) = D_1 - D_2 v_m + D_3 v_m^2 - D_4 v_m^3 + D_5 v_m^4 \qquad (13)$$

where values of J corresponding to v_m were calculated from the corresponding polystyrene calibrant molecular weight via Equation 9 using:

$$J(v_m) = (1.14 \times 10^{-4})(M_{PS}(v_m))^{1.72}$$

where we have used published values of K_{PS} and a_{PS} for polystyrene in THF at 25 deg C (45).

Stock solutions of samples were prepared with a known concentration (w/v) in THF in the range of 4×10^{-3} to 5×10^{-3} gm/ml. These stock solutions were filtered through a 0.22 micrometer Fluoropore filter prior to injection, and an injection size of 50 microliters was used. Of all the input parameters for MOLWT-II, a large potential source of error resides in the value used for the injected mass; loss of sample during prefiltration or adsorption on the SEC column packing can introduce significant error into the SEC/LALLS molecular weight data. Comparison of off-line and on-line $\overline{M}_w$ values is one check for full sample recovery, and this test was satisfied for the LB samples.

Mass recovery of MA samples was checked by using the concentration (DRI) detector response (mass/area ratio) of the corresponding LB arm; it was assumed that the detector response was identical for compositionally similar samples. Corrections for 38% and 9% sample loss were applied to the "mass injected" in the SEC/LALLS data for (SI-1) DVB and (SI-2) DVB, respectively.

Differential Refractometry (dn/dc). Stock solutions of polymer were prepared with known concentrations (w/v) in the solvent of choice, and the specific refractive index increment (dn/dc) was measured at 26 deg C with a KMX-16 Laser Differential Refractometer (LDC/Milton Roy). Sample concentrations typically were ca. 5×10^{-3} gm/ml.

Static Light Scattering. Off-line (static) values of the weight-average molecular weight ($\overline{M}_w$) were measured using solutions prepared with toluene, THF, and chloroform. Four or five solutions in the range 1.0×10^{-3} to 5.0×10^{-3} gm/ml for the LB and 0.1×10^{-3} to 0.5×10^{-3} gm/ml for MA samples were prepared via serial dilution of a stock solution which was prefiltered through a 0.22 micrometer Fluoropore filter. Also, a similar 0.22 micron filter was placed in the sample inlet line to the KMX-6 LALLS cell. The LALLS measurements were performed at 6-7 degrees forward scattering angle, and data were processed and plotted in the standard fashion as $Kc/\overline{R}_\theta$ vs. c (Equation 5); the intercept and slope of the best (visual) linear fit to the data gave the weight-average molecular weight ($\overline{M}_w$) and second virial coefficient (A_2), respectively.

Results

Off-Line $\overline{M}_w$ Measurements in Several Solvents. Table 1 shows results of dn/dc (column 3) and off-line $\overline{M}_w$ measurements (column 6) which were carried out in THF, toluene, and chloroform. The dn/dc also was calculated via Equation 1b using the weight fraction of each monomer (from proton NMR, "Experimental") and the dn/dc for the corresponding homo polymers. Values of the homopolymers in THF: styrene (0.190), isoprene (0.127), butadiene (0.132); toluene: styrene (0.108), isoprene (0.031), butadiene (0.032); chloroform: styrene (0.155), isoprene (0.093), butadiene (0.094). Values of dn/dc derived in this manner are presented in column 4.

Table I. Specific Refractive Index and Off-Line LALLS Data

Sample	Solvent	dn/dc (ml/gm) Meas.	Calc.	$\overline{M}_W \times 10^{-3}$ SEC/LALLS	$\overline{M}_W \times 10^{-3}$ Static LALLS	$A_2 \times 10^4$ (mol-cm³/gm²)
SB-1 (53% S)	THF	0.159	0.162	58.2	62.0	12.6
	Tol.	0.072	0.072	-	57.5	11.4
	Chlor.	0.124	0.126	-	58.1	13.2
(SB-1) DVB	THF	0.160		671	746	4.0
	Tol.	-		-	749	4.0
	Chlor.	-		-	892	4.7
K1107 (~16% S)	THF	0.137	0.137	154	163	9.9
	Tol.	0.041	0.043	130	194	2.9
	Chlor.	0.101	0.103	-	205	10.3
SI-1 (9% S)	THF	0.136	0.133	210	249	11.0
	Tol.	0.037	0.038	-	214	10.8
	Chlor.	0.101	0.099	-	222	10.4
(SI-1) DVB	THF	0.133		1770	-	-
SI-2 (23% S)	THF	0.140	0.141	143	138	8.8
	Tol.	0.049	0.048	-	135	10.0
	Chlor.	0.105	0.107	-	149	9.0
(SI-2) DVB	THF	-		1490	-	-
SI-3 (26% S)	THF	0.143	0.143	109	124	11.2
	Tol.	0.051	0.051	-	124	12.4
	Chlor.	0.108	0.109	-	115	11.9
(SI-3) DVB	THF	-		1110	-	-
SI-4 (48% S)	THF	0.155	0.157	59.4	64.6	11.2
	Tol.	0.068	0.068	-	60.0	11.4
	Chlor.	0.117	0.123	-	71.5	12.5
(SI-4) DVB	THF	-		870	1320	3.6
	Tol.	-			1250	2.8
	Chlor.	-			1300	3.1

For several of the DVB-linked multi-arm structures, it was impossible to obtain acceptable static LALLS data: solutions were extremely difficult to filter, and they exhibited noisy and unstable LALLS baselines.

The off-line measurements of the linear block copolymer "arm" samples were not difficult. However, in most cases, chloroform solutions demonstrated noticeably more LALLS baseline instability than those prepared in THF and toluene; intensity readings changed as much as 10% within several minutes regardless of the amount of solution prefiltration.

SEC Data. Tables I and II present data from the SEC/LALLS runs. Overall sample $\overline{M}_w$ values are given in Table I, while Table II shows $\overline{M}_w$ and polydispersity data for the major peak in each chromatogram along with the molecular weight of the "kill" polystyrene component in each LB sample. Table II includes results from both the SEC/LALLS and linear polystyrene calibration treatments; this table shows also the ratio (k) of the polystyrene-equivalent $\overline{M}_w$ to the value from SEC/LALLS for the major peak in the chromatogram. The eighth column in Table II gives the weight-average number of arms (f_w) for the MA samples, calculated from the $\overline{M}_w$ value of the major peak in the MA and LB chromatograms. In the case of the LB samples, the $\overline{M}_w$ from SEC/LALLS agrees favorably with that from off-line measurements. The approximately 19% higher off-line $\overline{M}_w$ obtained in toluene vs. THF for K1107 reflects aggregation; note the significantly lower A_2 in toluene.

In cases where static LALLS results were obtained for the DVB-linked samples, poor agreement was found with SEC/LALLS. In both cases shown in Table I ((SI-4) DVB) and ((SB-1) DVB), the SEC/LALLS $\overline{M}_w$ is considerably less than the off-line $\overline{M}_w$. The concentration detector (DRI) response showed no significant sample loss ("Experimental") following injection, and this discrepancy possibly results from breakup of sample aggregates during chromatography ("Discussion", below).

The SEC/LALLS chromatograms for LB samples K1107, SI-3, and SI-4 are shown in Figures 1, 2a, and 3a, respectively, with the chromatograms for (SI-3) DVB and (SI-4) DVB presented in Figures 2b and 3b, respectively. The corresponding log M(v) vs. v plots for (SI-3) DVB and (SI-4) DVB are given in Figures 2c and 3c, respectively, with representative values for the intrinsic viscosity ratio g'(v) included in the latter figures. The polystyrene calibration curve is included for comparison. Samples (SI-1) DVB and (SI-2) DVB showed sign inflections similar to (SI-3) DVB in the log M(v) vs. v plots, while (SB-1) DVB demonstrated behavior similar to (SI-4) DVB.

Table II. SEC Data

Sample	SEC/LALLS[1]		Polystyrene Calibration[2]			k[3]	f_w[4]
	$\overline{M}_w \times 10^{-3}$	$\overline{M}_w/\overline{M}_n$	$\overline{M}_w \times 10^{-3}$	$\overline{M}_w/\overline{M}_n$	$M_{KILL} \times 10^{-3}$		
SI-1	198	1.01	231	1.04	15.0	1.17	15.6
(SI-1) DVB	3.09×10^3	1.19	1.44×10^3	1.28	-	0.47	
SI-2	143	1.02	142	1.05	30.0	0.99	18.0
(SI-2) DVB	2.57×10^3	1.15	1.05×10^3	1.27		0.41	
SI-3	108	1.02	115	1.04	25.0	1.06	15.9
(SI-3) DVB	1.72×10^3	1.03	786	1.30		0.46	
SI-4	60.0	1.01	63.0	1.02	34.0	1.05	17.2
(SI-4) DVB	1.03×10^3	1.11	491	1.16		0.48	
SB-1	60.0	1.01	76.0	1.02	31.0	1.27	15.8
(SB-1) DVB	946	1.16	507	1.20		0.54	
K1107	166	1.01	175	1.03	13.6	1.05	-

(1) Calculated for the major polymer peak of the chromatogram.

(2) Calculated for the major polymer peak of the chromatogram. "M_{KILL}" is the peak mol. wt.

(3) Ratio of the polystyrene $\overline{M}_w$ (col. 4) to SEC/LALLS $\overline{M}_w$ (col. 2).

(4) Weight average branching functionality, using values in column 2.

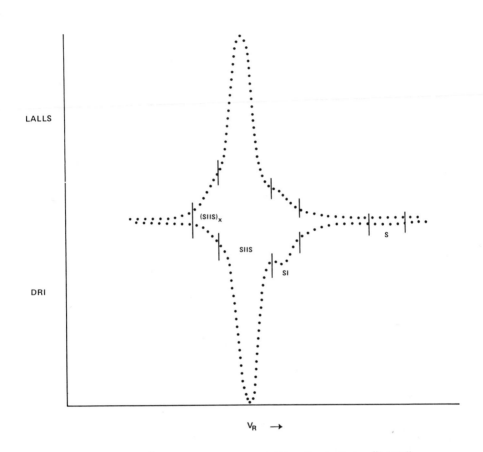

Figure 1. SEC/LALLA data for K1107. Peak S is "kill" polystyrene, while peaks SI, SIIS, and (SIIS)$_x$ are block copolymer, coupled block copolymer, and an unknown high molecular weight species, respectively. Values of $\overline{M}_w$ from SEC/LALLS are 8.26 x 10^4 and 2.79 x 10^5 for SI and (SIIS)$_x$.

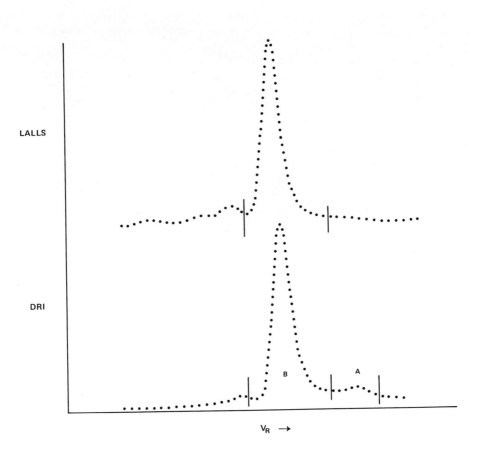

Figure 2a. SEC/LALLS chromatogram. Sample SI-3. Peak A is "kill" polystyrene; and peak B is LB.

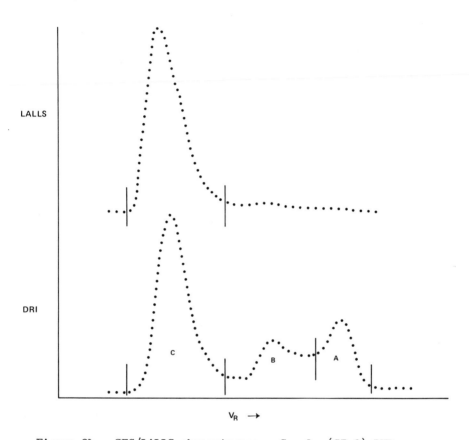

Figure 2b. SEC/LALLS chromatogram. Sample (SI-3) DVB.
Peak A is "kill" polystyrene; peak B is LB; and peak C is
DVB-linked MA.

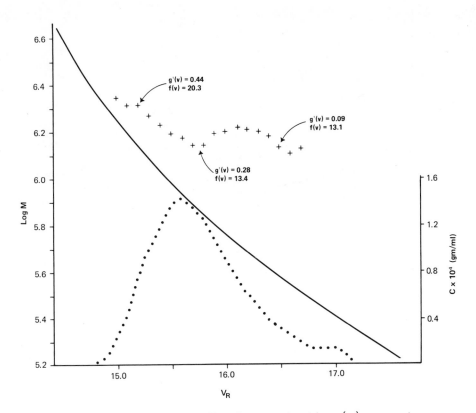

Figure 2c.   Plots of log M and concentration (c) vs. retention volume (v) for sample (SI-3) DVB. Values of the viscosity ratio g'(v) and branching functionality f(v) are
given for several retention volumes.

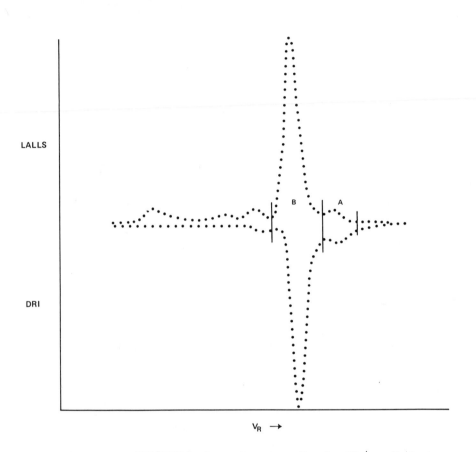

Figure 3a. SEC/LALLS chromatogram. Sample SI-4. Peak A is "kill" polystyrene; peak B is LB.

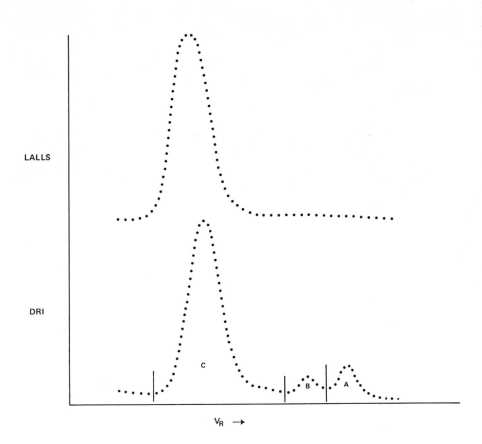

Figure 3b. SEC/LALLS chromatogram. Sample (SI-4) DVB.
Peak A is "kill" polystyrene; peak B is LB; and peak C is
DVB-linked MA.

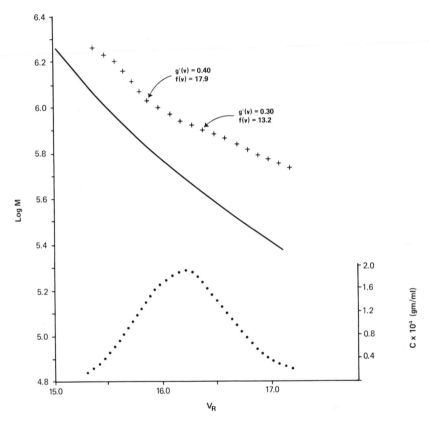

Figure 3c. Plots of log M and concentration (c) vs. retention volume (v) for sample (SI-4) DVB. Values of the viscosity ratio g'(v) and branching functionality f(v) are given for several retention volumes.

The chromatogram of Kraton 1107 shows the other components of the sample besides the major coupled diene S-I-I-S: small amounts of "kill" polystyrene, uncoupled S-I block copolymer, and material with higher molecular weight than that of SIIS are indicated. As indicated in Figures 2a and 3a, the LB polymers all showed a small polystyrene "kill" component and a high molecular weight shoulder on the block copolymer peak with a molecular weight of about twice that of the block copolymer.

The DVB-linked MA polymers showed evidence of "kill" polystyrene and block copolymer arm with peak elution volumes at the same position as in the LB chromatograms. Note that in Figure 2b, the LALLS chromatogram has a shoulder on the major peak which is not observable in the DRI chromatogram; this corresponds to a sign change in the slope of the log $M(v)$ vs. v relationship.

The most remarkable feature about the data from the LB and MA materials is the dramatic upward shift in the log $M(v)$ vs. v relationship which is induced by the multi-arm branching (Figures 2c and 3c). This is reflected by the change in k from values of ca. 1.0-1.3 to 0.5 for the LB and MA samples, respectively (Table II).

Discussion

Off-Line (Static Measurement and $\overline{M}_W$. Congruence of the off-line $\overline{M}_W$ measurement in the three solvents is consistent with compositional homogeneity of the LB arms ("Theory", above). Confirmation of the LB arm compositional uniformity was essential to the use of SEC/LALLS for investigation of the molecular weight/retention volume behavior of the MA polymers, since the dn/dc measured for the sample must correspond to that of polymer eluting in retention volume v. Agreement of the calculated and measured dn/dc (Table I) for the bulk sample is expected for block copolymers ("Introduction")). Off-line $\overline{M}_W$ measurement of MA samples were beset with experimental difficulties. For example, the (SI-4) DVB sample showed good $\overline{M}_W$ agreement in all three solvents, while (SB-1) DVB gave the same value in THF and toluene, but a significantly larger $\overline{M}_W$ in chloroform. Considerable LALLS baseline instability indicated aggregation/association behavior in the chloroform solutions of (SB-1) DVB.

The second virial coefficients (A_2) shown in Table I do not show significant dependence upon solvent for most of the LB samples. The only exception is K1107 which yielded an anomalously low A_2 value in toluene; this sample exhibited marginal solubility in toluene, and the low A_2 suggests unfavorable polymer-solvent interaction. The A_2 values for MA polymers, however, are consistently lower than their LB precursors. This decrease of A_2 with increased molecular weight (and branching) is consistent with dilute solution polymer theory (35).

SEC/LALLS Measurements: $\overline{M}_w$ Values. For most of the LB samples, SEC/LALLS $\overline{M}_w$ data are in reasonable agreement with off-line (static) values. This supports the validity of the SEC/LALLS experimental data since it shows that the calculated mass injected probably equals the eluting mass. Loss of mass via column adsorption, insoluble gel, etc., usually is manifested by a low value of the SEC/LALLS $\overline{M}_w$ relative to the off-line value. The off-line $\overline{M}_w$ values which were obtained for the MA samples were significantly larger than the SEC/LALLS values ((SI-4) DVB and (SB-1) DVB). The detector response characteristics of these MA samples, compared to the starting LB polymers, indicated no loss of mass during sample preparation and chromatography. A possible source of the $\overline{M}_w$ discrepancy is the presence of a small fraction of undissolved microgel in the stock solution along with a high proportion of loosely associated aggregates. The microgel probably is removed by filtration and column frits, while the aggregates would be dissociated in the strong shear fields in the flowing SEC solvent. The detector response data would not show sample loss from removal of a very small number of microgel particles and dissociation of the aggregated material. Such behavior is strongly suggested by the difficult filtration behavior of the MA sample, and the presence of microgel and aggregates will have a disproportionate effect on the LALLS response.

The differences in polystyrene-equivalent and absolute molecular weight for the LB and MA polymers are represented by the k values in Table II; they demonstrate dissimilarities in hydrodynamic volume/molecular weight which are conferred by monomeric composition and polymeric backbone structure. The decrease in k from ca. 1.0–1.3 for LB materials to about 0.5 for the corresponding MA reflects the much higher polymer segment density in the MA species.

The LB behavior can be compared with Tung's data (46) for styrene/butadiene block copolymers as well as with that author's presentation of Duc and Prud'homme's polystyrene/isoprene block copolymer data (47). The styrene/butadiene materials (45% styrene) exhibited polystyrene-equivalent molecular weights about a factor of 1.35 larger than their true molecular weight. Although he did not carry out experiments with block copolymers of polystyrene/isoprene, Tung did study homopolymers of isoprene and butadiene; these data suggest that the polystyrene-equivalent molecular weight of styrene/isoprene block copolymers would be closer to the true value than in the case of styrene/butadiene. The k value for SB-1 (1.27) agrees with Tung's data for styrene/butadiene block copolymers of similar composition, and the lower values (1.0–1.17) for the styrene/isoprene block copolymers are consistent with Tung's homopolymer results.

The weight-average branching functionality (f_w) for the major MA peak (Table II) falls between 16 and 18 for all the samples. This probably reflects the constancy of experimental preparation conditions: temperature, concentration, solvent, and the ratio [DVB]/[RLi].

The dramatic upward shift in the MA molecular weight/elution volume curve was accompanied, in several of the samples, by changes in the sign of the slope. Figure 2c illustrates this. Such a sign in inflection was found in the molecular weight/elution volume behavior of samples (SI-1) DVB, (SI-2) DVB, and (SI-3) DVB, and it is manifested by appreciably smaller $\bar{M}_w/\bar{M}_n$ for SEC/LALLS compared with polystyrene calibration data (Table II). The polystyrene calibration forces any chromatogram to yield identical elution volume/molecular weight characteristics. (The SEC/LALLS method typically gives a slightly lower polydispersity than linear polystyrene calibration, due to detector response differences and opposite effects of column/hardware band spreading on the molecular weight calculation (28, 29, 31). However these effects generally are small relative to the polydispersity differences shown here.) The SEC/LALLS data for the lowest molecular weight MA samples, (SI-4) DVB and (SB-1) DVB, showed continually decreasing molecular weight with retention volume and a smaller slope than the polystyrene calibration. The latter difference contributes to the smaller polydispersity ($\bar{M}_w\bar{M}_n$) given by SEC/LALLS. A sign inflection was noted in the SEC/LALLS molecular weight/elution volume behavior of cellulose tricarbanilate (48), and these workers have ascribed it to "branching"; the qualitative rationale is that branched polymer with a high segment density can elute after a less-branched polymer with a larger hydrodynamic volume and smaller molecular weight. Such SEC/LALLS data for the cellulose derivatives and some of the MA samples in this work may reflect distributions of branching structure. Further studies are necessary to elucidate this.

The representative viscosity ratio ($g'(v)$) values shown in Figures 2c and 3c reflect considerably higher segment density of the MA species relative to their linear homolog of identical molecular weight. However, the variation of $g'(v)$ with M is contrary to that expected from theory, which predicts an increase in this parameter with decreasing molecular weight (21). The variation of $g'(v)$ with M is qualitatively predictable from comparison of the polystyrene calibration curve and data shown in Figures 2c and 3c: Equation 12 shows that if $M_{PS}(v)/M_{MA}(v)$ decreases with molecular weight, as with data shown here, $g'(v)$ must decrease. This result casts doubt on the validity of the universal calibration as well as assumption of a constant ratio (k) for M_{PS}/M_{LB}. The latter certainly may be in error, and further work with LB materials of constant composition and a range of molecular weights would be necessary to evaluate the appropriateness of the assumption. Also, the SEC behavior of the MA materials may not conform to the universal calibration curve; the work of Bi and Fetters (1) on similar samples containing butadiene-styrene and iso-prene-styrene block copolymer arms showed that the universal calibration method gave erroneous molecular weight informa-tion for high molecular weight samples. The structure which Bi and Fetters ascribe to the DVB-linked block copolymers, i.e., an inner "core" poly(diene) surrounded by a "shell" of polystyrene, might confer unusual SEC elution behavior which is related to their finding that the intrinsic viscosity depends on arm length and not branching functionality. (Similar MA materials prepared with homopolymer arms of isoprene or butadiene gave accurate molecular weights via the universal calibration (8, 15).)

Finally, chromatography artifacts may contribute to failure of universal calibration. At high sample concentra-tions, so-called "viscous streaming" (43) retards SEC elution of high molecular weight polymers. This may in fact account for the inflection in the molecular weight/elution volume behavior shown by the three higher molecular weight MA samples: this viscosity/concentration effect might be selec-tively retarding elution of some higher molecular weight species within the same sample. The effect may be operating to a lesser extent in the SEC behavior of the two lower molecular weight MA samples. In the latter case, it will tend to decrease the slope of the log M/V_R curve.

In summary, the approach outlined here is a straight-forward method for determining representative values of viscosity ratios $[\eta]_{MA}/[\eta]_{LB}$; certainly g' values significantly less than 1.0 are expected for such highly branched polymers (33). However, the anomalous dependence of g'(v) on M_{MA} suggests that 1) the core/shell hydrodynamic configuration and/or chromatographic artifacts invalidate universal calibration, and/or 2) the LB elution behavior does not conform to that of polystyrene in the assumed, constant manner. Further work is necessary to elucidate these points.

Literature Cited

1. Bi, L. K.; Fetters, L. J. Macromolecules 1976, 9, 732.
2. Von Meerwall, E.; Tomich, D. H.; Hadjichristidis N.; Fetters, L. J. Macromolecules 1982, 15, 1157.
3. Raju, V. R.; Menezes, E. V.; Marin, G.; Graessley, W. W.; Fetters, L. J. Macromolecules, 1981, 14, 1668.

4. Morton, M.; Helminiak, T. E.; Gadkary, S. D.; Bueche, F. J. Polym. Sci., 1962, 57, 471.
5. Zelinski, R. P.; Wofford, C. F. J. Polym. Sci., Part A, 1965, 3, 93.
6. Roovers, J. L.; Bywater, S. Macromolecules, 1972, 5, 385.
7. Roovers, J. L.; Bywater, S. Macromolecules, 1974, 7, 443.
8. Hadjichristidis, N.; Guyot, A.; Fetters, L. J. Macromolecules, 1978, 11, 668.
9. Hadjichristidis, N.; Fetters, L. J. Macromolecules, 1980, 13, 191.
10. Decker, D.; Rempp, P. C. R. Acad. Sci., Ser. C., 1965, 261, 1977.
11. Worsfold, D. J.; Zilliox, J. G.; Rempp, P. Canad. J. Chem., 1969, 47, 3379.
12. Bi, L. K.; Fetters, L. J. Macromolecules, 1975, 8, 90.
13. Kohler, A.; Polacek, J.; Koessler, T.; Zilliox, J. G.; Rempp, P. Eur. Polym. J., 1972, 8, 627.
14. Zilliox, J. G. Makromol. Chem., 1972, 156, 121.
15. Quack, G.; Fetters, L. J.; Hadjichristidis, N.; Young, R. N. Ind. Eng. Chem. Prod. Res. Dev., 1980, 19, 587.
16. Martin, M. K.; Ward, T. C.; McGrath, J. E. in "Anionic Polymerization"; McGrath, J. E., Ed.; ACS SYMPOSIUM SERIES No. 166, American Chemical Society: Washington, D.C., 1981; p. 558.
17. Roovers, J.; Hadjichristidis, N.; Fetters, L. J. Macromolecules, 1983, 16, 214.
18. Hadjichristidis, N.; Roovers, J. J. Polym. Sci. (Polym. Phys. Ed.), 1974, 12, 2521.
19. Bauer, B. J.; Hadjichristidis, N.; Fetters, L. J.; Roovers, J. L. J. Am. Chem. Soc., 1980, 102, 2410.
20. Benoit, H. J. Polym. Sci., 1953, 11, 507.
21. Zimm, B. H.; Stockmayer, W. H. J. Chem. Phys., 1949, 17, 1301.
22. Candau, F.; Rempp, P.; Benoit, H. Macromolecules, 1972, 5, 627.
23. McCrackin, F. L.; Mazur, J. Macromolecules, 1981, 14, 1214.
24. Stockmayer, W. H.; Fixman, M. Ann. N. Y. Acad. Sci., 1953, 57, 334.
25. Roovers, J. Polymer, 1979, 20, 843.
26. Roovers, J.; Toporowski, P. M. Macromolecules, 1981, 14, 1174.
27. Ouano, A. C.; Kaye, W. J. Polym. Sci. (Polym. Chem. Ed.), 1974, 12, 1151.
28. McConnel, M. L. Am. Lab., 1978, 10 (5), 63.
29. Jordan, R. C. J. Liquid Chromatog., 1980, 3, 439.
30. Hamielec, A. E.; Ouano, A. C.; Nebenzahl, L. L. J. Liquid Chromatog., 1978, 1, 527.

31. Jordan, R. C.; McConnel, M. L. in "Size Exclusion Chromatography (GPC)"; Provder, T., Ed.; ACS SYMPOSIUM SERIES No. 138, American Chemical Society: Washington, D.C., 1979; pp. 107–129.
32. Axelson, D. E.; Knapp, W. C. J. Appl. Polym. Sci., 1980, 25, 119.
33. Agarwal, S. H.; Jenkins, R. F.; Porter, R. S. J. Appl. Polym. Sci., 1982, 27, 113.
34. Benoit, H.; Froelich, D. in "Light Scattering from Polymer Solutions"; Huglin, M. B., Ed.; Academic: New York, 1972; p. 468.
35. Flory, P. J. in "Principles of Polymer Chemistry"; Cornell University Press: Ithaca, New York, 1953.
36. Stacey, K. A. in "Light Scattering in Physical Chemistry"; Academic Press: New York, 1956.
37. Tanford, C. in "Physical Chemistry of Macromolecules"; John Wiley and Sons: New York, 1961.
38. Kaye, W. Anal. Chem., 1973, 45, 221A.
39. Kaye, W.; Havlik, A. J. Appl. Opt., 1973, 12, 541.
40. Huglin, M. B. in "Light Scattering from Polymer Solutions"; Huglin, M. B., Ed.; Academic: New York, 1972; p. 165.
41. Schulz, G. V.; Wunderlich, W.; Kirste, R. Makromol. Chem., 1964, 75, 22.
42. Machtle, W.; Fischer, H. Angew. Makromol. Chem., 1969, 7, 147.
43. Yau, W. W.; Kirkland, J. J.; Bly, D. D. in "Modern Size Exclusion Chromatography"; John Wiley and Sons: New York, 1979.
44. Grubisic, Z.; Rempp, P.; Benoit, H. J. Polym. Sci., 1967, B 5, 753.
45. Hellman, M. Y. in "Liquid Chromatography of Polymers and Related Materials"; Cazes, J., Ed.; Marcel Dekker: New York, 1977; p. 31.
46. Tung, L. H. J. Appl. Polym. Sci., 1979, 24, 953.
47. Ho-Duc, N.; Prud'homme, J. Macromolecules, 1973, 6, 472.
48. Cael, J. J.; Cannon, R. E.; Diggs, A. O. in "Solution Properties of Polysaccharides"; Brant, D. A., Ed.; ACS SYMPOSIUM SERIES No. 150, American Chemical Society: Washington, D.C., 1980; p. 43.

RECEIVED December 20, 1983

Determination of Thermoset Resin Cross-link Architecture by Gel Permeation Chromatography

A. J. AYORINDE, C. H. LEE, and D. C. TIMM
University of Nebraska, Lincoln, NE 68588–0126

W. D. HUMPHREY
Brunswick Corporation, Lincoln, NE 68504

Gel permeation chromatography is the method of choice for analysis of thermoplastic resin systems. Corrected for imperfect resolution, chromatogram interpretation yields accurate molecular descriptions, including theoretical, kinetic distributions (1,2). The current research is designed to extend the utility of this analytical tool to the analysis of thermoset resins.

Kinetic mechanisms (3) are such that low molecular weight species are present in a cured resin; in fact, the molar concentration of dimers, trimers, etc. usually exceeds that for higher molecular weight species. An exception is a Poisson distribution, but oligomeric species are still abundant. If a cured thermoset resin is prepared such that a large surface area to volume ratio is achieved, solvent leaching provides an effective method for sample preparation. Analysis of extracts (4,5) provides data descriptive of monomeric content and oligomeric, molecular distributions. Such extracts contain definitive information with respect to the extent of cure as well as a description of the crosslink architecture. Average molecular weights between crosslink sites plus crosslink density within the insoluble, resin fraction can be determined.

Observations for cured epoxy resins and resins derived from 1,2-polybutadiene crosslinked with t-butylstyrene are reported. These resins find applications in aerospace industry, including high performance, Kevlar 49, filament wound, pressure vessels on Skylab and the Space Shuttle.

Population Density Distributions

Chain-growth polymerization. A 1,2-polybutadiene polymer is crosslinked with t-butylstyrene, utilizing a free radical initiator. Reaction rates include

Initiation	$I \rightarrow 2A_o$	k_i
Propagation	$A_j + M^o \rightarrow A_{j+1}$	k_p
Branching	$A_j + P_m \rightarrow A_{j+m}$	k_b
Termination	$A_j + A_k \rightarrow P_{j+k}$	k_t

The 1,2-polybutadiene initially formulated is a commercially available material supplied by Colorado Specialty Company and Nippon Soda. An anionic polymerization, initiated by a butyl lithium, is likely used in its manufacture. This results in a molar distribution of constitutive molecules defined by a Poisson distribution for batch polymerizations. Thus, the number and weight average molecular weights are nearly equal. The current research further assumes that this distribution is sufficiently narrow such that all polybutadiene molecules are of the same molecular weight, which is described by the degree of polymerization n. This constraint greatly simplifies the mathematical description to be developed for the population of molecules during the subsequent chain-growth cure initiated by dicumyl peroxide.

Fisher (6), in a discussion of relative rates of reaction, states that the styrenic free radical is more likely to react with a styrene molecule than with the polyunsaturated 1,2-polybutadiene. The relative rates are expected to differ by orders in magnitude. Therefore, the propagation reaction rate is expressed in terms of the molecularly mobile monomer, t-butylstyrene. The consequence is that the 1,2-polybutadiene will be crosslinked primarily by t-butylstyrene segments.

The extracts from a quality resin contain oligomeric molecules of a degree of polymerization less than that for the 1,2-polybutadiene. These species are a consequence of simultaneous propagation and termination reactions. Their population density distribution is also descriptive of that portion of molecules which react with polymeric species, initially forming a branched, and later a crosslinked, structure within the resin. Research shows that the average molecular weight of the oligomeric fraction correlates with the crosslink average molecular weight within the insoluble, crosslinked resin fraction (7). Such is a kinetic consequence of the competition between branching and termination reactions in the above reaction model.

For free radical species of degree of polymerization less than that for the 1,2-polybutadiene used in the formulation, a kinetic reaction analysis results in the following relationships expressed in terms of the molar concentration of primary free radicals A_o.

$$\frac{dA_o}{dt} = 0 = 2k_i I - (k_p M + k_b P_{TOT} + k_t A_{TOT}) A_o$$

$$A_o = 2k_i I / (k_p M + k_b P_{TOT} + k_t A_{TOT})$$

$$\frac{dA_1}{dt} = 0 = k_p MA_o - (k_p M + k_b P_{TOT} + k_t A_{TOT})A_1$$

$$A_1 = A_o(k_p M)^1/(k_p M + k_b P_{TOT} + k_t A_{TOT})^1 = \frac{A_o}{D^1}$$

The cumulative molar concentrations of polymeric and activated intermediates are P_{TOT} and A_{TOT}, respectively. The denominator is $D = (k_p M + k_b P_{TOT} + k_t A_{TOT})/k_p M$. The analysis recognizes that these activated intermediate species must be saturated and, therefore, do not experience generation through branching/crosslinking reactions which normally require unsaturation. For primary free radicals A_o, conservation of population includes initiation, propagation, branch formation and termination reactions. The latter is assumed to be by combination. For free radicals that contain monomer segments j, $0 < j < n$, the initiation rate is superseded by a propagation rate. The rate of accumulation or depletion within the batch reactor is negligible for these activated intermediates. These expressions are representative of a recurring type relationship.

For molecules at a degree of polymerization n or larger, the mathematical model incorporates branch formation reactions which include a free radical of size j and a polymeric specie of degree of polymerization $m \geq n$. The consequence is the formation of a free radical of molecular size $j + m$. Furthermore, due to the relatively high concentration initially of the 1,2-polybutadiene constituent at $j = n$, the derivation assumes that all polymeric species of size $j \geq n$ are unsaturated and are capable of branch and/or crosslink formation. Polymeric species are denoted by P_j; free radical intermediates are described by A_j. Therefore, the first activated intermediate capable of formation by branching reactions is A_n via $A_o + P_n \rightarrow A_n$. Conservation laws yield

$$\frac{dA_n}{dt} = 0 = k_p MA_{n-1} - (k_p M + k_b P_{TOT} + k_t A_{TOT})A_n + k_b A_o P_n$$

$$A_n = A_o/D^n + k_b A_o P_n/k_p MD$$

As the degree of polymerization increases, all possible combinations of reactions forming a free radical via branching must be considered. Thus

$$\frac{dA_{n+1}}{dt} = 0 = k_p MA_n - (k_p M + k_b P_{TOT} + k_t A_{TOT})A_{n+1} + k_b(A_o P_{n+1} + A_1 P_n)$$

Previous expressions for A_n and A_1 can be substituted, yielding

$$A_{n+1} = \frac{A_o}{D^{n+1}} + \frac{k_b A_o}{k_p MD}(P_{n+1} + \frac{2P_n}{D})$$

The factor 2 is a consequence of the second term in the expression for A_n and the term $k_b A_1 P_n$ in the conservation expression dA_{n+1}/dt. At a degree of polymerization $j = n + 2$

$$\frac{dA_{n+2}}{dt} = 0 = k_p M A_{n+1} - (k_p M + k_t A_{TOT} + k_b P_{TOT}) A_{n+2} + k_b (A_o P_{n+2} + A_1 P_{n+1} + A_2 P_n)$$

Substitution of the several expressions for A_{n+1}, A_1, A_2 and a collection of similar terms yields

$$A_{n+2} = \frac{A_o}{D^{n+2}} + \frac{k_b A_o}{k_p MD}\left(P_{n+2} + \frac{2P_{n+1}}{D} + \frac{3P_n}{D^2}\right)$$

This type of recurring formula represents the molar concentration of free radicals up to a degree of polymerization $j=2n-1$. At molecular weights twice that of the initial 1,2-polybutadiene, $j=2n$, the initial substitution of the expression for A_n in the rate of formation due to branching occurs and results in a second major change in the overall functionality of the descriptive relationship for the concentration of activated intermediates. Consider the conservation laws at this degree of polymerization

$$\frac{dA_{2n}}{dt} = 0 = k_p M A_{2n-1} - (k_p M + k_b P_{TOT} + k_t A_{TOT}) A_{2n} + k_b \{A_o P_{2n} + A_1 P_{2n-1} + \cdots$$
$$+ A_{n-1} P_{n+1} + A_n P_n\}$$

Solving this expression for A_{2n} after expressing A_j, $1 \leq j \leq n$ and A_{2n-1} in terms of A_o yields

$$A_{2n} = \frac{A_o}{D^{2n}} + \frac{k_b A_o}{k_p MD}\left(P_{2n} + \frac{2P_{2n-1}}{D} + \frac{3P_{2n-2}}{D^2} + \cdots + \frac{(n+1)P_n}{D^n}\right)$$
$$+ \left(\frac{k_b}{k_p MD}\right)^2 A_o (P_n P_n)$$

Thus, the addition of a third function occurs for the first time at $j=2n$. Continuation of the derivation will result in a series of rather complex functionality, but one which will be mathematically defined.

The degree of polymerization intervals of interest are, therefore, comprised of distinct regions determined by the initial molecular weight of the 1,2-polybutadiene, n.

$$j < n \qquad A_j = A_o/D^j$$

$$n \leq j < 2n \qquad A_j = A_o/D^j + \frac{k_b A_o}{k_p MD} \sum_{k=1}^{j-n+1} k P_{j+1-k}/D^{k-1}$$

$$2n \leq j \qquad A_j = A_o/D^j + \frac{k_b A_o}{k_p MD} \sum_{k=1}^{j-n+1} kP_{j+1-k}/D^{k-1}$$

$$+ (\frac{k_b}{k_p MD})^2 A_o \sum_{m=n}^{m=j-n} \frac{(m+1-n)(m+2-n)}{2} \sum_{k=n}^{k=j-m} P_k P_{j-k-m+n}/D^{m-n} \qquad (1)$$

A set of first order differential equations descriptive of the molar concentrations for polymeric species is given. Species, less than size n, are saturated and, therefore, accumulate only within the batch reactor and do not participate in branch/cross-link reactions. Molecules greater in size than n are unsaturated and will experience the reaction described by branch formation. As the extent of cure progresses, this reaction forms chain networks within the resin. Representative equations are

$$j < n \qquad \frac{dP_j}{dt} = A_o^2 k_t (j+1)/2D^j$$

$$n \leq j < 2n \qquad \frac{dP_j}{dt} = A_o^2 k_t (j+1)/2D^j + \frac{A_o^2 k_t k_b}{2k_p MD} \sum_{k=0}^{j-1} \frac{(k+1)(k+2)}{D^k} P_{j-k} \qquad (2)$$

An expression for macromolecules greater than twice that of the initial 1,2-polybutadiene will necessarily be more complex due to the last equation of Expression (1).

Experimentally, macromolecules greater than 2n are usually crosslinked to an extent that they are essentially insoluble, being attached to the network resin structure. The intent of the model is to explain comparative observations in oligomeric population density distributions obtained through analysis of extracts of thermoset resins. The model clearly shows that for activated intermediates A_j and for polymeric species P_j frequency distributions are comprised of additive functions for distinct regions in molecular weight. The functionality at a lower degree of polymerization is contained within the distribution at a larger degree of polymerization, relative to the size of the initial polybutadiene component.

The term D^j can be expressed in terms of the relative rates of branch formation plus termination compared to propagation. Since this relative rate is numerically small, a truncated series of ln(1+x) results in the valid approximation

$$D^{-j} = \exp(- \frac{k_b P_{TOT} + k_t A_{TOT}}{k_p M} j) \qquad (3)$$

The argument of the exponential is x. Therefore, molar distributions of oligomeric species leached from cured resins will be presented on semilogarithmic graphs.

Data observed for cured resins are presented by Figure 1.
Formulations and cures were identical except for the molecular
weight of the 1,2-polybutadiene. The functionality of the oligomeric fraction leached from cured resins clearly shows expected
dependency of the population density distribution on molecular
weight of the original polymer. Numerical chromatogram analysis
(1) corrects for imperfect resolution. The calibration utilized
a set of linear, epoxy resins formed from the step-growth polymerization of nadic methyl anhydride and phenyl glycidyl ether. The
molecular distributions of these materials are a Poisson (8) distribution of different average molecular weight. The calibration
and subsequent interpretation has been extensively tested using
thermoplastics (9,10) of known, kinetic distribution. However,
the effects of hydrodynamic volume on molecular weight on the present nonlinear oligomeric fraction of varying chemical composition
is unknown. Thus, the assignment of degree of polymerization is
on a relative basis.
    Figure 1 is graphed consistent with the functionality of
Equations 2. The degree of polymerization 329 is a constant of
calibration. The 450-1200 molecular weight, 80% 1,2-polybutadiene
resin has an initial inflection point at about 700 molecular
weight, a second at 2,570 molecular weight. Equations 1 and 2
predict such, though the second break point is somewhat greater
than the simplified model predicts. Integration and Trommsdorff
(11) effects are expected to influence precise locations. The extracts were leached from quality cured resins being evaluated for
aerospace applications. The initial break points for the 2,000
molecular weight, 80% 1,2-polybutadiene resin and for the 4,400
molecular weight, 70% 1,2-polybutadiene resin show expected dependency on molecular weight (see Table I). The initial and second
break points for the 3,000 molecular weight, 90% 1,2-polybutadiene
and the second break point for the 2,000 molecular weight specimens are absent, primarily due to low oligomeric resin content at
expected degrees of polymerization. Less than one percent of the
former resin is soluble. The second break points for the two,
highest molecular weight specimens are in the regions of insoluble, crosslinked structures. Observations are tabulated, see
Table I.

Step-growth polymerization. Epoxy resins were prepared from
nadic methyl anhydride and Epon 828. This bifunctional oxirane
also supplies reactive hydrogen sites. The major component is at
i=0, minor components include oligomers with i=1,2,3. Their concentrations rapidly diminish as degree of polymerization increases.

$$\begin{array}{ccc} & C & C \\ C-CC(O\phi C\phi OCCC)_i O\phi C\phi OCC-C \\ O & C \quad OH & C \quad O \end{array}$$

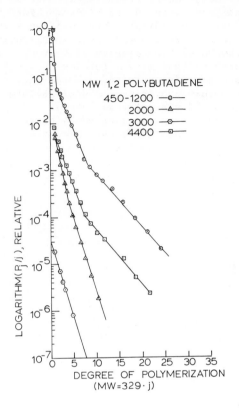

Figure 1. Oligomeric frequency distribution, chain-growth polymerization.

Table I: Hydrocarbon Resin Formulation

$\overline{MW}$	1,2-polybutadiene Parts	% 1,2	t-butyl-styrene Parts	Dicumyl Peroxide Parts	Figure 1 Break Point 1st	$\overline{MW}$ 2nd
450-1200	70	80	30	1.9	660	2570
2000	80	80	20	1.9	1410	--
3000	80	90	20	1.9	--	--
4400	80	70	30	1.9	2500	--

Cure at 24 hours, 140°C

Basic industrial catalysts were utilized. The ionic, polymerization mechanism (12) results in the reactive hydrogen site alternately reacting with the anhydride and oxirane groups. The chemistry of the reaction is such that an alcohol end group reacts
with monomeric anhydride, forming an acid group. This group will
then react with an oxirane group, forming an alcohol. Thus, the
reactive hydrogen site is conserved. The molecule's backbone
structure will contain anhydride residuals alternating with oxirane residuals coupled via ester linkages. The molecule will also
contain pendent side chains terminated by oxirane groups. Acid
end groups on one molecule will, therefore, react with monomeric or
oligomeric species containing the oxirane functionality. Both
reactions result in the coupling of two molecules

$$P_j + P_k \rightarrow P_{j+k} \qquad\qquad k_c$$

If one of the species is monomeric oxirane, then $j = 1$. Likewise, if one of the polymeric species supplied the oxirane, then
$j > 1$. The molecule with the acid group is at degree of polymerization k. The degree of polymerization indexes the number of
oxirane residuals within the macromolecule. Though the reaction
sequence is simplified, it retains the essence of one molecule
reacting with every other molecule. This step-growth mechanism
(13) develops the thermoset resin microstructure.

For batch polymerizations initially void of polymeric species, the molar distribution of polymeric species is expressed by

$$P_j(\tau) = P_1(\tau) \qquad \{1-\exp(-\tau)\}^{j-1}$$

An excellent approximation for small values of $\exp(-\tau)$ is

$$P_j(\tau) = P_1(\tau)\exp(-\exp(-\tau)(j-1)) \qquad\qquad (4)$$

The time variable τ is the eigenzeit transform

$$\tau = \int_0^t k_c P_{TOT}(t)dt$$

The first moment of the distribution is P_{TOT}, the total, cumulative molar concentration of polymeric material. As the molecular
weight of polymeric species increases, branching and crosslinking
reactions yield a thermoset resin. Chromatography analysis of
epoxy resin extracts confirms the expected population density distribution described by Equation 4, as is shown in Figure 2. Formulations and cure cycles appear in Table II.

Three of the four resins yield extracts of the functionality
of Equation 4. The slope of the exponential decay allows for the
evaluation of τ. The resin, see Table II, initiated by benzyl dimethyl amine (BDMA) at the stated cure cycle, when subjected to
leaching yields an extract of low solubility and a distribution
of oligomeric molecules of low number average molecular weight.

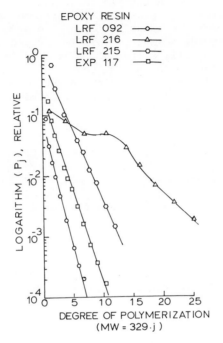

Figure 2. Oligomeric frequency distribution, step-growth polymerization.

Table II: Epoxy Resin Formulation

Resin	Anhydride	Epoxy	Catalyst	HDT, °F
LRF-216	NMA	Epon 828	ATC 3	107
*LRF-215	Polyoxy-propyleneamine	DER 332	---	139
EXP-117	NMA	Epon 828	EMI-24	211
LRF-092	NMA	Epon 828	BDMA	276

Anhydride Cure   12 hrs., 125°F; 2 hrs., 225°F; 4 hrs., 290°F
* Amine Cure   16 hrs., 80°F; 2 hrs., 150°F

Such a resin is highly crosslinked. The resin Exp 117 is more soluble and the oligomeric species are of a greater average molecular weight. Thus, the average extent of crosslink development has diminished. The resin LRF 215 is an amine cured epoxy; thus, its relative placement will depend, in part, on the specific refractive index increment contribution of the amine component compared to the anhydride segment. This is not expected for this resin to be substantially different. The slope of the population distribution of oligomeric species contained in the resin's extract is of a still higher molecular weight. Thus, the extent of crosslinked development is less than the previous two resins discussed. The system LRF 216 is the least cured of the four resins.

The emphasis of the current research is on molecular structure of oligomeric fractions leached from quality cured, industrial resins. However, the potential for applications in quality control should not be overlooked. Chromatography analysis provides positive feedback capable of molecular descriptions of extent of cure actually achieved. Oligomeric distributions coupled to kinetic reaction analysis allows for detailed estimates of crosslink architecture within the resin (7).

Molecular Architecture

For quality cured thermoset resins, approximately one percent of the mass is soluble when subjected to long-term leaching with tetrahydrofuran. Equilibrium is approached in two weeks; resin swell is not visually noticeable. The monomeric, chemical structures are such that the hydrocarbon resins exhibit more pronounced viscoelastic properties; whereas, the epoxy resins are similar to elastic bodies when subjected to tensile testing at room temperature. Therein, LRF 216 is less sensitive to flaws and is more nonlinear in tensile or compressive stress-strain analysis.

Data in Table II for the epoxy resin sets are ordered according to increasing extent of crosslink development. Heat distortion temperatures are an indication of molecular weight between crosslink sites. The average degree of polymerization of the soluble oligomeric fraction reported was obtained by gel permeation chromatography. In conjunction with Figure 2, results show that as the average molecular weight of the oligomeric fraction diminishes and as the resin becomes less soluble, the number average molecular weight between crosslinks decreases and the crosslink density increases within the insoluble network fraction. Similar data for the hydrocarbon resins have been reported (14).

Acknowledgments

Financial and technical support from the Brunswick Corporation and the Engineergin Research Center are appreciated.

Nomenclature		Units
A_j	concentration of free radical, degree of polymerization j	moles/volume
A_{TOT}	$=\sum_{j=0}^{\infty} A_j$, cumulative molar distribution	moles/volume
D	$= (k_pM + k_tA_{TOT} + k_bP_{TOT})/k_pM$	
j	degree of polymerization, $j = 0,1,2,3\ldots$	
k	degree of polymerization, $k = 0,1,2,3\ldots$	
k_b	rate constant, branch formation	vol/mole time
k_c	rate constant, molecular combination	vol/mole time
k_i	rate constant, initiation	vol/mole time
k_p	rate constant, propagation	vol/mole time
k_t	rate constant, termination	vol/mole time
M	monomer concentration	moles/volume
m	degree of polymerization, $m = n, n+1, n+2\ldots$	
n	degree of polymerization of 1,2-polybutadiene	
P_j	concentration of polymeric specie of degree of polymerization j	moles/volume
P_{TOT}	$=\sum_{1=n} P_k$ for hydrocarbon resin	moles/volume
P_{TOT}	$=\sum_{x=1} P_k$ for epoxy resin	
t		time

Literature Cited

1. Timm, D.C.; Rachow, J.W., J. Poly. Sci. 1975, 13, 1401.
2. Pickett, H.E.; Cantow, M.J.R; Johnson, J.F., J. Poly. Sci. 1968, C-21, 23, 67.
3. Flory, P.J. "Principles of Polymer Chemistry"; Cornell University Press: Ithaca, NY, 1953.
4. Yen, H.C.; Tien, C.S.; Timm, D.C., Proc. 2nd World Congress of Chemical Engineering 1981, VI, 381.
5. Plass N.C.; Timm, D.C.; Liu, S.H.; Humphrey, W.D., Proc. Conference International du Caoutchouc 1982, I, 1.
6. Fisher, J.P., Angew. Makromol. Chem. 1973, 33, 35.
7. Timm, D.C.; Ayorinde, A.J.; Huber, F.K.; Lee, C.H., submitted to International Rubber Conf. '84, Moscow.
8. Adesanya, B.A.; Yen, H.C.; Timm, D.C.; Plass, N.C., ACS symposium "Recent Advances in Size Exclusion Chromatography", in press.
9. Timm, D.C.; Kubicek, L.F., Chem. Engr. Sci. 1974, 29, 2145.
10. Scamehorn, J.F.; Timm, D.C., J. Poly. Sci. 1975, 13, 1241.
11. Trommsdorff, E.; Kohle, H.; Legarrly, Pl, Makromol. Chem. 1947, 1, 169.
12. Feltzin, J., Am. Chem. Soc. Mtg. Chicago 1964, No. 40.
13. Yen, H.C., M.S. Thesis, University of Nebraska-Lincoln, Lincoln, NE, 1981.
14. Humphrey, W.D.; Liu, S.H.; Timm, D.C.; Plass, N.C., Proc. 6th Conférence Européenne des Plastiques 1982, II, 28.

RECEIVED September 12, 1983

Size Exclusion Chromatography Analysis of Epoxy Resin Cure Kinetics

GARY L. HAGNAUER and PETER J. PEARCE[1]

Polymer Research Division, Army Materials & Mechanics Research Center, Watertown, MA 02172

Liquid size exclusion chromatography (SEC) is applied to investigate the isothermal cure kinetics of the reaction between pure N,N'-tetraglycidyl methylene dianiline (TGMDA) and 4,4'-diaminodiphenyl sulfone (DDS) monomers over the temperature range 121° to 187°C. Intermediate reaction products are isolated by preparative SEC, identified and used as standards for SEC calibration. Monomer and soluble reaction product concentrations, molecular weight averages, and gel content are monitored as functions of reaction time by analytical SEC. A 3rd order rate expression describing the early stages of cure is established and Arrhenius relationships describing the temperature dependence of the rate constant and the onset of gelation are determined. Reaction mechanisms are discussed and the effects of variations in stoichiometry of TGMDA/DDS resins on the network structure and properties of the cured resin are considered.

Epoxy resins containing N,N'-tetraglycidyl methylene dianiline (TGMDA) and 4,4-diaminodiphenyl sulfone (DDS) are widely used in the manufacture of fiber-reinforced structural composites for aircraft. However, the commercial resin formulations are generally quite complex and may include several different types of epoxy resins, additional curing agents, catalysts, organic solvents, and additives to facilitate processing or modify properties of the cured resin. An accurate assessment of the cure kinetics is virtually impossible since the resins are often partially reacted or "staged" during their formulation and "prepregging" which generates a host of ill-defined, intermediate reaction products and because a variety of reactions which proceed at different rates and by different mechanisms may occur during cure. Indeed impurities

[1]Current address: Materials Research Laboratories, Ascot Vale, Victoria, Australia

and synthesis by-products present in commercial TDMDA are found to
have a significant effect on curing behavior when DDS alone is added
as the curing agent (1). To begin to understand the curing behavior
of the commercial resins it is essential first to investigate and
understand the curing chemistry of simpler model systems. Preliminary studies have shown that it is possible to accurately monitor
cure kinetics and, at least during the early stages of cure, to
elucidate the curing chemistry if pure TGMDA and DDS monomers are
used (2).

In this paper liquid size exclusion chromatography (SEC) is
applied to investigate the isothermal cure kinetics of the reaction
between pure TGMDA and DDS monomers over the temperature range 121°
to 187°C. The objective is to gain a better understanding of the
epoxy resin curing chemistry and to evaluate the temperature
dependence of the curing reaction. Intermediate reaction products
are isolated by preparative SEC, identified and used as standards
for SEC calibration. Monomer and soluble reaction product concentrations, molecular weight averages and gel content are monitored
as functions of reaction time by SEC. From stoichiometric studies
a rate expression describing the early stages of cure is developed
and an Arrhenius relationship is determined from the temperature
dependence of the rate constant.

Experimental

Preparative liquid chromatography techniques were applied to purify
the TGMDA monomer (1). The monomer used for this study is a pale
yellow liquid with a viscosity of approximately 1300 centipoise at
50°C and an epoxy equivalent weight (EEW) of 108g/eq. The theoretical EEW of the TGMDA monomer is 105.5g/eq. The curing agent DDS is
a white, crystalline (mp, 162°C) powder and was highly pure
(approx., 99%) as received from Aldrich Chemical Co. TGMDA/DDS
resin formulations were prepared by heating the weighed components
to 90°C and then mixing to form homogeneous solutions (approx.,
30g). Except during sampling the resin formulations were stored in
sealed containers at -13°C. Chromatographic and spectroscopic
analyses showed that no reaction occurred during mixing and that
upon storage the formulations remained unreacted for at least 6
months.

A Perkin-Elmer DSC 1B instrument was used to study the isothermal
cure (polymerization) behavior of the resin formulations. Samples
(5-10mg) were weighed in aluminum DSC sample pans on a microbalance
and transferred to the DSC heating stage. The temperature of the
heating stage was preset at the curing temperature and the cures were
conducted in a nitrogen atmosphere. About 10-20 samples per resin
formulation were partially cured over a range of time intervals. The
reactions were terminated by rapidly lowering the temperature and
transferring the sample pans to 25 mL volumetric flasks and adding

tetrahydrofuran (THF). To facilitate dissolution the flasks were
agitated and the samples allowed to soak for 1-4 days. Except for
through a 0.2 μM Millipore membrane filter in preparation for SEC
analysis. Only soluble components were analyzed by SEC.

A Waters Associates ALC/GPC-244 instrument with M6000A solvent
delivery system, M720 system controller, M730 data module, 710B WISP
auto-injector and M440 UV detector was used for the SEC analyses and
operated under the following conditions:

Column Set: μStyragel (2 x 500Å, 3 x 100Å)
Sample Concentration: 0.2-0.5 μg/μL
Injection Volume: 20-60 μL
Mobile Phase: THF (UV grade, Burdick & Jackson Labs)
Flow Rate: 1 mL/min
Detector: UV 254nm
Run Time: 45 min

A Waters Associates Prep LC System/500 was used for preparative
SEC. Samples were injected using a 12 mL loop valve and the column
set consisted of two 2.5-in diameter x 4-ft length columns with 80-
100Å and 700/2000Å Styragel packing. Operating conditions are shown
below:

Sample Concentration: 20g/100mL
Mobile Phase: THF (UV grade, Burdick & Jackson Labs)
Flow Rate: 40mL/min
Detector: differential refractive index (RI)
Run Time: 94 min

Cure Mechanism

Epoxy-amine curing reactions are known to be exceedingly complex.
More than one reaction can occur and the temperature dependence of
each reaction may be quite different. For the TGMDA-DDS system,
moisture and resin impurities can not only behave as catalysts but
also may affect the network structure and properties of the cured
resin (2). Because of the tetra-functional nature of the monomers,
steric effects may lead to alternative reactions and highly crosslinked network structures may occur relatively early in the curing
reaction. Indeed as polymerization proceeds, viscosity increases
and the glass transition temperature of the reaction mixture
gradually approaches and may exceed the curing temperature. As a
result, reactive species become diffusion limited and eventually may
either seek other reaction pathways or stop reacting entirely.

Denoting the structures

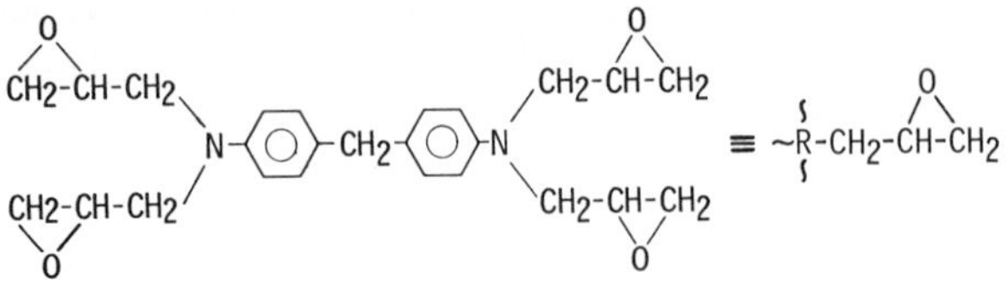

TGMDA

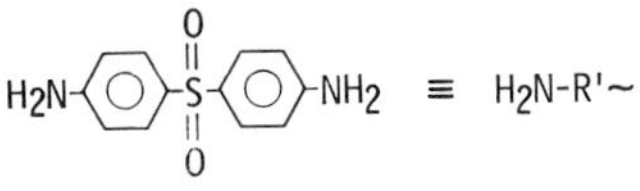

DDS

the most likely reactions and reaction product structures are shown below -

(1)   epoxy-primary amine addition

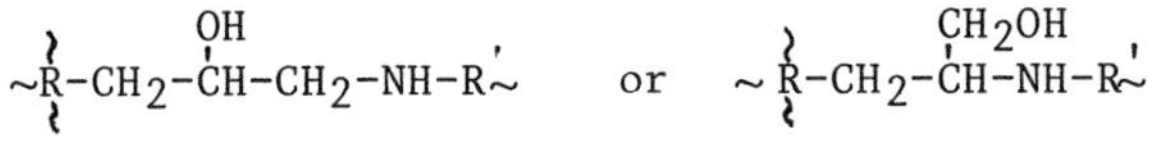

$$\sim\overset{\displaystyle\wr}{\underset{\displaystyle\wr}{R}}-CH_2-\overset{\displaystyle OH}{\underset{\displaystyle |}{CH}}-CH_2-NH-R\sim \quad , \quad or \quad \sim\overset{\displaystyle\wr}{\underset{\displaystyle\wr}{R}}-CH_2-\overset{\displaystyle CH_2OH}{\underset{\displaystyle |}{CH}}-NH-R\sim \quad ,$$

(2)   epoxy-secondary amine addition

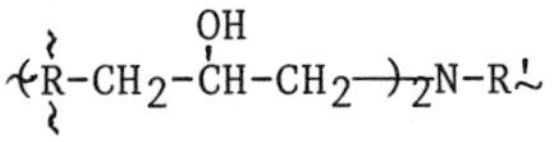

$$\left(\overset{\displaystyle\wr}{\underset{\displaystyle\wr}{R}}-CH_2-\overset{\displaystyle OH}{\underset{\displaystyle |}{CH}}-CH_2\right)_2 N-R\sim \qquad and\ isomers$$

(3)   epoxy-hydroxyl addition

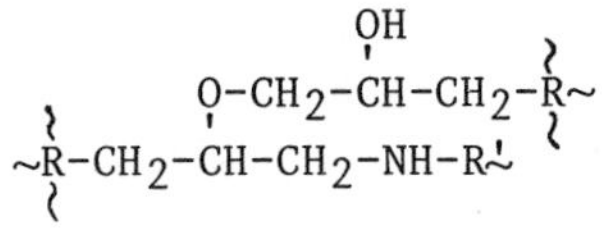

$$\underset{\displaystyle\wr}{\overset{\displaystyle O-CH_2-\overset{OH}{\overset{|}{CH}}-CH_2-\overset{\wr}{\underset{\wr}{R}}\sim}{\sim\overset{\displaystyle\wr}{\underset{\displaystyle\wr}{R}}-CH_2-CH-CH_2-NH-R\sim}}$$

plus isomers and other products from the reaction of species from (1) and (2)

(4)   epoxy-epoxy homopolymerization

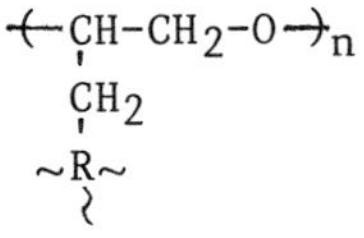

$$\left(\overset{\displaystyle CH-CH_2-O\right)_n}{\underset{\displaystyle\underset{\displaystyle\sim R\sim}{CH_2}}{|}}$$

Preparative SEC

Samples for preparative SEC were prepared by partially polymerizing about 10g of a TGMDA/DDS mixture. To optimize the formation of the initial, relatively simple reaction products, a TGMDA resin form- ulation consisting of 25% by weight DDS was cured for 23 minutes at 145°C in vacuo. The RI detector trace from the preparative SEC of this reaction mixture is illustrated in Figure 1. Four injections were made successively at 60 minute intervals and fractions 1:1 and 2:1 were collected as indicated. Following the application of a vacuum to remove solvent, approximately 0.3g fraction 1:1 and 0.1g fraction 2:1 were realized. Analytical SEC and reverse phase high performance liquid chromatography (HPLC) showed that fraction 1:1 was 97% pure. Using Fourier transform infrared (FTIR) and H^1/C^{13} NMR spectroscopy, fraction 1:1 was identified as the TGMDA–DDS epoxy-primary amine addition product shown below.

Fraction 1:1

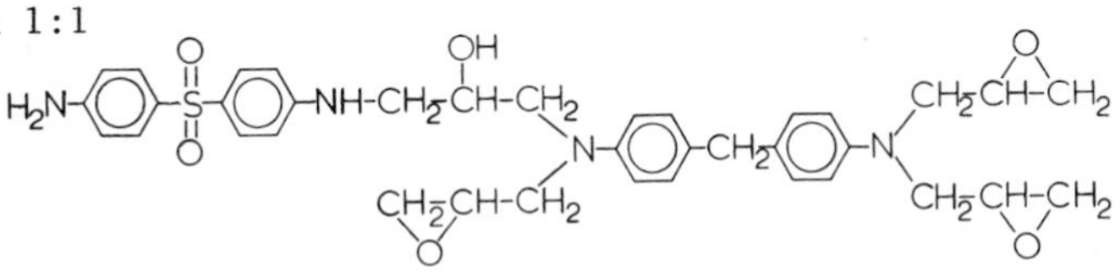

1-1 PRODUCT

Although a single peak was observed using SEC, fraction 2:1 was found to contain three components by HPLC analysis. The SEC analysis indicates that the components of fraction 2:1 have quite similar molar volumes. To ascertain the ratio of monomers in each component and obtain more information about the reaction mechanism, FTIR spectra of the 2:1 components were run and compared with the spectra of TGMDA, DDS, and the 1-1 product. Absorbance bands for hydroxy ($3500 \ cm^{-1}$) and secondary amine ($3410 \ cm^{-1}$) groups were apparent in the spectra of all three components. There was no evidence suggesting the presence of aliphatic ether linkages of the type $-CH_2-O-CH_2-$ or $>CH-O-CH_2-$. From the results of the chromato- graphic and FTIR analyses, the following structures are postulated for the components found in fraction 2:1.

Fraction 2:1

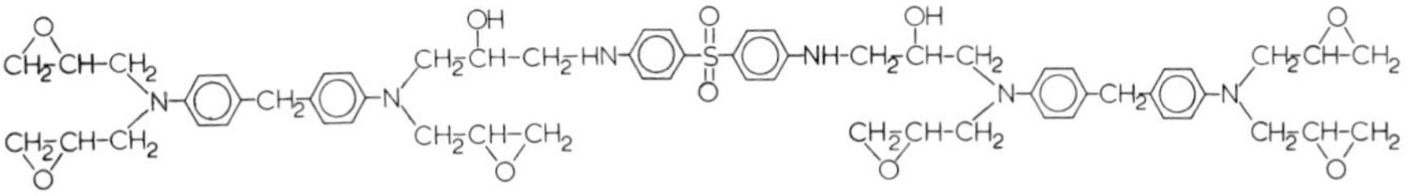

2-1 PRODUCT

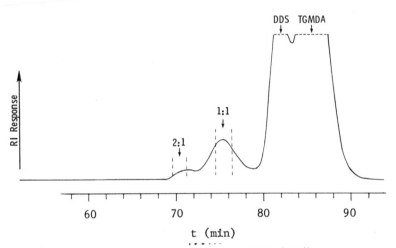

Figure 1. Preparative SEC of TGMDA/DDS (25%) resin
reacted 23 min at 145 °C.

1-2 PRODUCTS

Fractions of the higher molecular weight, more complex reaction products were obtained by the preparative SEC of further advanced TGMDA/DDS reaction mixtures. For example, the proposed components of the next highest oligomer fraction are the 3-1, 2-2, and 1-3 TGMDA-DDS products. The relative ratio of the products depends upon the initial composition of the TGMDA/DDS resin formulation.

Analytical SEC

With standards for calibration, SEC may be applied to determine monomer and reaction product concentrations, molecular weight (MW) averages, and gel content in TGMDA/DDS reaction mixtures. Typical SEC chromatograms are shown in Figure 2. The chromatograms are displaced along the ordinate to illustrate changes in composition accompanying the cure of the TGMDA/DDS(25%) resin at 177°C. Reaction products elute with retention times between 28 and 32 minutes. Areas under the SEC peaks and peak segments are directly proportional to the concentrations of the components. The initial reaction products are TGMDA-DDS oligomers. As the reaction proceeds, higher MW, soluble products are formed. The onset of gelation is indicated by the formation of insoluble products. At the onset of gelation (30 min in Figure 2), substantial amounts of the monomers remain unreacted. As gelation continues, areas of peaks representative of high MW products rapidly diminish and eventually the concentrations of extractable monomers approach zero.

The weight percentage of each component or set of components designated C_i may be calculated from their respective peak areas A_i

$$\%W_i = \frac{A_i K_i}{C_o V} \cdot 100\% \tag{1}$$

where $K_i = C_{i,s} \cdot V_s / A_{i,s}$ is the calibration constant, C_o is the concentration ($\mu g/\mu L$) of the sample assuming complete solubility, V is the injection volume (μL), and subscript "s" denotes the respective parameters for the calibration standards. Using a 254nm

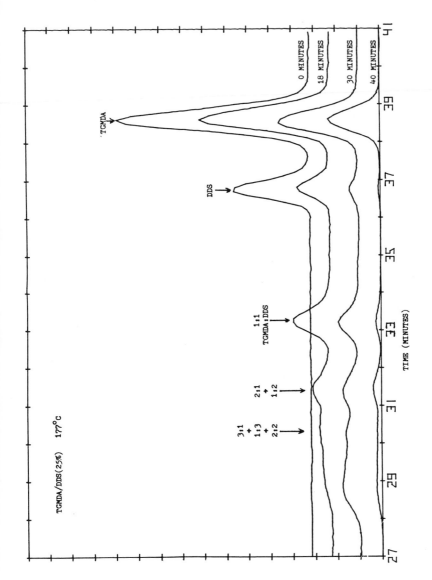

Figure 2. SEC analysis of TGMDA/DDS (25%) resin cured for different periods of time at 177 °C.

UV absorbance detector, it is noted that the calibration constants for TGMDA and DDS are quite similar and that the 1-1 product has a different constant which is essentially identical to those of the higher MW products. For components having retention times less than 30 min, their total weight percentage may be calculated from the sum of area segments A_j between 28 and 32 min taken at 0.1 min intervals; i.e.,

$$\%W_i = \frac{K \sum A_j}{C_o \cdot V} \cdot 100\% \qquad (2)$$

where K is a constant. The gel content is calculated using the equation

$$\%gel = 100\% - \sum \%W_i \qquad (3)$$

The MW calibration curve is shown in Figure 3. Discrete MW values obtained by averaging the MW's of components eluting at the same retention times are indicated as data points. Even though DDS has the lowest MW, its retention time is less than that of TGMDA. This apparent anomaly is attributed to differences in the extent of solvation of the two molecules. DDS has amino-groups which are highly polar and may hydrogen bond with THF to form a solvated species having a larger molar volume that of TGMDA in THF. Data points for the intermediate MW reaction products fit on the same line as DDS and may be extrapolated (dashed line) to account for higher MW components.

Standard equations are applied to calculate number-, weight-, and z-average MWs

$$M_n = \sum W_i / \sum (W_i / M_i) \qquad (4)$$

$$M_w = \sum (W_i M_i) / \sum W_i \qquad (5)$$

$$M_z = \sum (W_i M_i^2) / \sum (W_i M_i) \qquad (6)$$

where W_i is the weight fraction of component(s) C_i of average molecular weight M_i. For components eluting in the extrapolated region at time t_i, M_i is defined by

$$\log_{10} M_i = 6.697 - 0.1172 \cdot t_i \qquad (7)$$

Results and Discussion

The isothermal cure kinetics of a series of TGMDA/DDS resin formulations were investigated over the temperature range 121°–187^oC. Figure 4 illustrates data obtained for the resin TGMDA/DDS(25%) at 177^oC. During the early stage of cure prior to the onset of

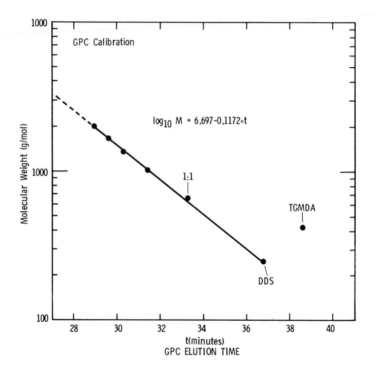

Figure 3. SEC calibration plot for TGMDA-DDS reaction products.

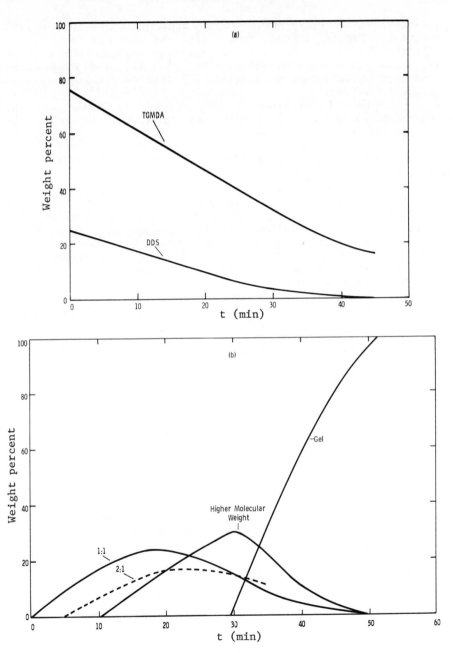

Figure 4. Weight percentages (a) of TGMDA and DDS and (b) of TGMDA/DDS reaction products vs. reaction time for TGMDA/DDS (25%) cured at 177 °C.

gelation, simple epoxy-primary amine addition is the predominant reaction. The 1-1 product forms first and increases steadily in concentration until its rate of reaction exceeds its rate of formation. The total concentration of higher MW products approaches a maximum at the onset of gelation and then decreases sharply as gelation proceeds.

A plot of experimental data (Figure 5) shows that the concentrations C (mol/kg) of TGMDA and DDS decrease in parallel as the reaction time increases; i.e.,

$$-d[TGMDA]/dt = -d[DDS]/dt \qquad (8)$$

Equation 8 holds over nearly 20% of the total theoretical extent of reaction and adequately describes the early stages of reaction of the TGMDA/DDS(25%) resin over the entire temperature range investigated (Figure 6). Indeed FTIR spectroscopic analysis of the reaction mixture at various cure times supports the conclusion that there is a one-to-one correlation between epoxide concentration and TGMDA concentration and that no major side reactions occur during the early stages of cure.

Stoichiometric studies show that the reaction is first-order with respect to the concentration of TGMDA and second-order with respect to DDS in the early stages of reaction.

$$-d[TGMDA]/dt = k_3[TGMDA] \ [DDS]^2 \qquad (9)$$

Results from rate studies at 161° and 177°C are shown in Table I. Third-order rate constants k_3 calculated from data obtained at 161°C are in excellent agreement over a broad range of TGMDA/DDS resin compositions. The slight increase in the 177°C k_3 values with increasing DDS concentration is attributed to problems in sampling with the DSC heating stage; i.e., at higher temperatures and higher DDS concentrations, the initial rate of reaction is sufficiently large that the sample heat-up time becomes a significant factor in rate determinations (Table I).

The third-order rate expression (Equation 9) is applicable over the temperature range 121° to 187°C. The Arrhenius relationship describing the temperature dependence of the rate constant k_3 (Figure 7) is

$$k_3 \ [kg^2mol^{-2}min^{-1}] = 2.15 \times 10^6 . exp(-16600/RT) \qquad (10)$$

where $R = 1.9872$ cal·mol^{-1}K^{-1}, T is temperature (°K), and the activation energy is 16,600 cal·mol^{-1}.

An activation energy may also be determined from the gelation data (Figure 8). For example, the relationship

$$t_{gel}^{-1} \ [min^{-1}] = 2.22 \times 10^6 . exp(-16100/RT) \qquad (11)$$

was determined from the temperature dependence of the reaction time

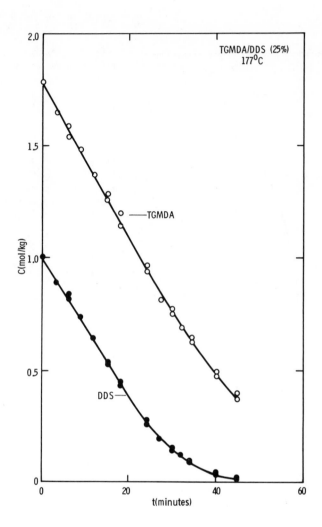

Figure 5.   TGMDA and DDS concentrations vs. reaction time
for TGMDA/DDS (25%) cured at 177 °C.

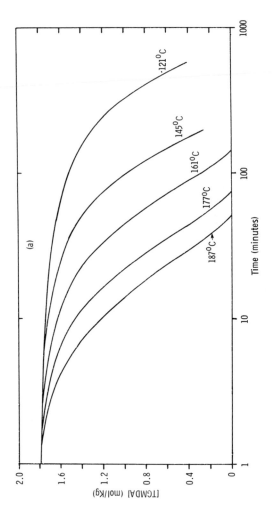

Figure 6. Semilogarithm plots of (a) TGMDA and (b) DDS monomer concentrations vs. reaction time for TGMDA/DDS (25%) cured at different temperatures.

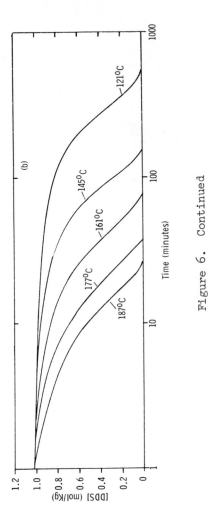

Figure 6.   Continued

Table I

TGMDA/DDS Rate Study

Temperature °C	DDS (weight-%)	[DDS]$_o$ (mol·kg^{-1})	[TGMDA]$_o$	$-d[\text{TGMDA}]/dt$ (mol·kg^{-1}min^{-1})	k_3 (kg^{2}mol^{-2}min^{-1})
161	15	0.605	2.014	0.0078	0.0106
	25	1.008	1.778	0.0195	0.0108
	37	1.49	1.493	0.0355	0.0107
	50	2.02	1.185	0.0526	0.0109
177	15	0.605	2.014	0.0132	0.0179
	25	1.008	1.778	0.0332	0.0184
	37	1.49	1.493	0.0741	0.0224

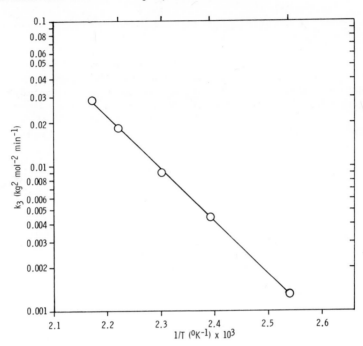

Figure 7. Arrhenius plot of the TGMDA/DDS third-order rate constant k_3.

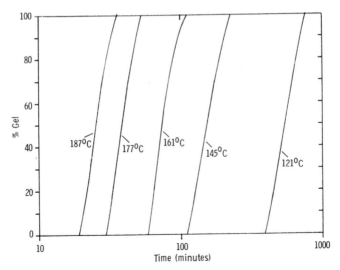

Figure 8. Semilogarithm plot of weight percentage gel vs. reaction time for TGMDA/DDS (25%) cured at different temperatures.

to the onset of gelation t_{gel} for the TGMDA/DDS(25%) resin formulation (Figure 9). It is noted that the activation energy 16,100 cal.mol^{-1} is quite similar to the value derived from kinetics data for the initial epoxy-primary amine addition reaction.

The rate of change and values of the MW parameters provide information relating to the formation of the gel network. MW parameters for the TGMDA/DDS(25%) resin are plotted versus reaction time at 177°C (Figure 10). M_w and M_z are most sensitive to the formation of high MW products and approach infinity (indicated by dashed lines) as the reaction nears the onset of gelation. The finite values of the parameters beyond the onset of gelation and the downward curvature of the plots (solid lines) are a consequence of the fact that only soluble components can be analyzed by SEC and that the highest MW products tend to be incorporated into the gel network first. Also, the M_z plot tends to curve downward earlier because the SEC calibration is no longer applicable in the high MW region when the reaction reaches the stage where a variety of highly branched products are formed. Only M_n can be interpreted beyond the onset of gelation by including the weight fraction of insoluble gel in the numerator of Equation 4.

The MW parameters and gel formation are dependent upon stoichiometry. Plots of the number-average MW ratio $(M_n)_t/(M_n)_o$ and gel fraction f_{gel} versus the fraction of TGMDA reacted at time t (Figure 11) show that, as the weight % DDS is decreased, less TGMDA is required to react for the mixture to attain specific degree of polymerization and gel fraction values. The data suggests that the effective functionality of DDS is less than 4. Although not prevalent in the early stage of cure, the epoxy-hydroxyl addition reaction would effectively increase the functionality of TGMDA and produce a similar result. The plots would overlap only if the functional groups of each monomer were equally reactive. Consequently, a more highly crosslinked network is formed as the % DDS is decreased, at least down to concentrations of 15% DDS. At lower DDS concentrations or higher extents of reaction, the trend may be reversed as other, perhaps more complex, reactions occur.

Recently, FTIR spectroscopy studies have been reported which support the above observations. Moacanin et al (3) concluded that two reactions dominate the TGMDA/DDS cure: epoxy-primary amine addition is the principal reaction occurring during the early stage of cure followed by the epoxy-hydroxyl addition reaction. Indeed they find that the rate of epoxy-hydroxyl addition is at least an order of magnitude slower than for the epoxy-primary amine reaction at 177°C. Furthermore, Morgan et al (4) report that the epoxy-secondary amine addition and epoxy-epoxy homopolymerization reactions also occur at 177°C but at rates that are approximately 10 and 200 times slower, respectively, than the epoxy-primary amine reaction.

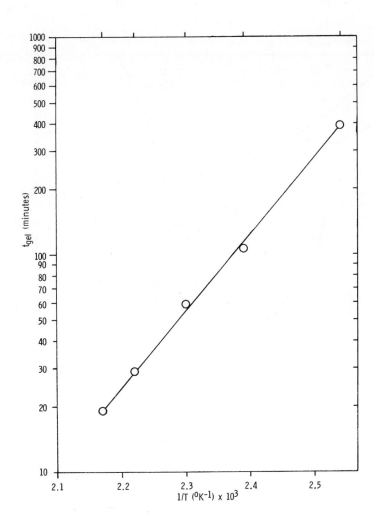

Figure 9.   Arrhenius plot of the reaction time to the onset of gelation t$_{gel}$ for the TGMDA/DDS (25%) resin.

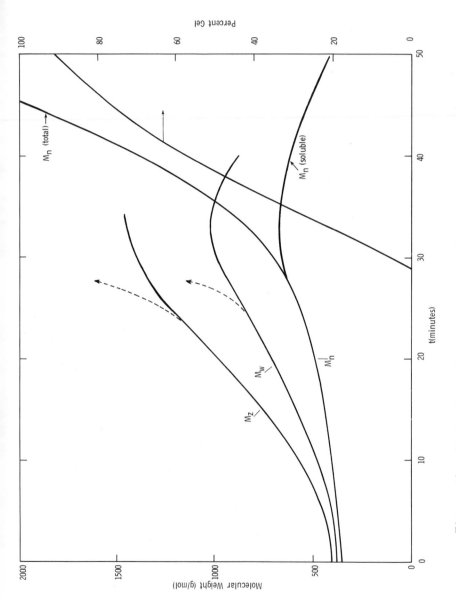

Figure 10. Plot of the MW parameters and the weight percentage gel vs. reaction time for TGMDA/DDS (25%) cured at 25 °C.

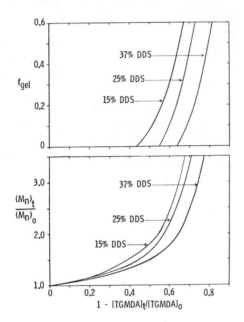

Figure 11. Gel fraction f_{gel} and number-average MW ratio (degree of polymerization) vs. fraction TGMDA reacted at time t for TGMDA/DDS mixtures cured at 177 °C.

Conclusions

1. Preparative and analytical SEC are powerful techniques for investigating the curing behavior of epoxy resins.
2. Epoxy-primary amine addition is the only reaction detectable in the earliest stage of the TGMDA/DDS cure and is the predominant reaction at least up to the onset of gelation.
3. TGMDA/DDS cure kinetics is adequately described by a third-order rate expression (Equation 9) during the early stage of cure.
4. The same rate expression and cure mechanism for the early stage of cure apply over the temperature range 121° to $187^{\circ}C$. Arrhenius relationships for the temperature dependence of the rate constant and onset of gelation have been determined.
5. Variations in the degree of polymerization and gel fraction data with changes in stoichiometry suggest the DDS secondary amine is not as reactive as the primary amine and beyond the early stages of reaction indicate the presence of alternative reaction mechanisms. Consequently, it is expected that variations in stoichiometry of TGMDA/DDS resins should have a predictable effect on network structure and properties of the cured resin.

Acknowledgment

The authors are grateful to Ms. Judith Brodkin for her assistance in running the SEC experiments.

Literature Cited

1. Hagnauer, G. L. and Pearce, P. J. Org. Coat. Appl. Polym. Sci. Proc., Am. Chem. Soc.,1982, 46, 580.
2. Hagnauer, G. L.; Pearce, P. J.; LaLiberte, B. R. and Roylance, M. E. Org. Coat. Appl. Polym. Sci. Proc., Am. Chem. Soc., 1982, 47, 429.
3. Moacanin, J.; Cizmecioglu, M.; Tsay, F. and Gupta, A. Org. Coat. Appl. Polym. Sci. Proc., Am. Chem. Soc., 1982, 47, 587.
4. Morgan, R. J., Happe, J. A. and Mones, E. T., Lawrence Livermore National Laboratory. Preprint UCRL-88513, presented at the 28th National SAMPE Symposium, Anaheim, CA, April 1983.

RECEIVED October 13, 1983

Sulfonated Poly(styrene–Divinylbenzene) Networks

Scission Study Using Aqueous Size Exclusion Chromatography

DAVID H. FREEMAN and XUN LIANG[1]

Department of Chemistry, University of Maryland, College Park, MD 20742

Several reports have been made on the degradation by hydrogen peroxide (Fenton's reagent) of ion exchange resins (sulfonated PSDVB). The reaction causes weight loss, swelling and eventual dissolution (1). Diffusion and secondary chemical reactions are possible; the scission rates vary oppositely with the amount of crosslinking (2). The rate differs among the DVB isomers, meta or para, used in the original PSDVB copolymerization (3).

Few structural tools are available to assess the structure of crosslinked networks directly. A frequent approach is to derive such information from kinetic study of unreacted monomer during the polymerization process (4). The possibilities for deriving structural information by characterizing network fragments has not been fully explored.

The groundwork for the present study has been developed in previous studies. For example, scission through peroxide oxidation or ultrasonic treatment of polystyrene chains has been found by size exclusion chromatography to involve preferential attack at the mid-chain position (5). Given this evidence it is expected that branched polymers should give a correspondingly skewed molecular weight distribution. This reasoning suggests one of the pathways by which a scission experiment may convey topological information.

PSDVB copolymers and their ion exchange derivatives consist of a three dimensional four-connected network structure. Such networks may have a statistically isotropic structure that includes tetrahedral cells, such as the "X" unit structure described by Flory (6). The four-connectedness results from the expected pairwise chain connecting function of

[1] On leave from Department of Chemistry, Nankai University, Tianjin, People's Republic of China

0097–6156/84/0245–0355$06.00/0

the DVB units. However, a more complete topological
model for such networks should, at least in principle,
provide for the 34 configurations described by Ziabecki
(7). Although it is important to consider the
theoretical perspectives, experiments that probe
polymer molecular topology are rare indeed.

If one assumes that a network consists only of
Flory tetrahedral cells, or X-units, the average mass
of the unit cell can be estimated from the monomers
used in the reaction mixture. Consider a reaction
mixture that incorporates f moles of a sum, D, of meta
and para DVB isomers plus an assumed equal portion E,
of the usual meta and para isomers of the EVD
(ethylvinylbenzene) contaminants, plus 1-f moles of
styrene. The estimated X-unit contains an average of
(1-f)/f moles of styrene per mole of DVB. The mass of
the average X-unit, MX, can be calculated from the
following expression:

$$MX = ((1-f)/f)MS + ME + MD \qquad (1)$$

where M refers to mass and S, E, and D refer to the
incorporated moities from styrene, plus the assumed
equal mole fractions, f, of DVB and EVB monomers,
respectively. The mass of the sulfonated X-unit in the
cation exchange derivative, is obtained by modifying
the values of MS, ME and MD by adding the appropriate
sulfonate and counterion masses.

The average mass of a single chain between
crosslinks, MC is estimated from:

$$MC = (MX - MD)/2 \qquad (2)$$

The division by two denotes the topological requirement
(6) of two inter-crosslink chains per DVB, in the
present model of a closed X-type network structure.
(Consider two X units with their chain ends joined
together. There are four chain lines and two
vertices.)

An example of the possibility that network
scission experiments may be subject to topological
interpretation is suggested by the results reported by
Hookway and Shelton (2). Of particular interest is the
degelation point where the network dissolves.
(Degelation implies transition through a gel point that
may or may not be related structurally to the usual
non-gel to gel transition observed in the corresponding
network synthesis. The data (ref. 2, Fig. 3) show that
hydrogen peroxide causes the release of about 0.5 mole

of carbon dioxide per crosslink to reach the degelation point. Since there are two lengths of chains per DVB vertex, this corresponds to about one mole of carbon dioxide released for each intercrosslink chain in the original network. This suggests the possibility that the scission reaction may be topologically selective and it may be of value for investigating the topology of fragment formation, and for studying the chemistry of scission degradation.

The goal of the present work is to examine the feasibility of obtaining topologically significant results from scission experiments, and to determine whether the topology of branched structures can be studied using topologically selective scission processes. (It has not yet been proved that any reaction offers such selectivity.) The first step, as will be described, is to examine the high points of the molecular weight distribution of the scission fragments.

The present experimental approach is based on the chromatographic advantages provided by the diol or glycerol derivatives of porous silica stationary phases available for use in HPLC. These have recently become available for estimating the molecular size of polyelectrolytes using aqueous size exclusion chromatography. The conditions for reproducible polyelectrolyte size measurements, and their possible purturbations have been summarized by Barth (8).

EXPERIMENTAL

Bio-Rad AG50W resins (sulfonated PSDVB), 50-100 mesh, were treated with NaOH and distilled water washes. The weighing state was obtained after 12 hours of drying in air at 75-80 C. Reagent grade chemicals were used throughout.

The scission reaction was carried out with a fixed addition of 1.50g of the dry resin, 10 mg of ferrous sulfate heptahydrate and 50 ml of 3% w/v hydrogen peroxide in a round Pyrex flask. The evolved carbon dioxide was vented to the atmosphere through serial traps containing sulfuric acid followed by a soda lime sorption tube. The magnetically stirred reaction flask was submerged in an oil bath heated with an immersed electrical coil and a magnetic stirrer positioned below the bath. The temperature was maintained at 50 +/- 1 C. After varied times 1.0 ml samples of liquid were withdrawn. There were fewer than six withdrawals in a given reaction sequence.

The liquid chromatographic analysis was carried
out using serial 4x300mm u-Bondagel E-125 and E-500
columns obtained from Waters Associates, Inc. The
carrier was prepared to contain (A) 0.25M sodium
perchlorate, 0.1% sodium lauryl sulfate that was
dissolved and brought to pH 7.2 using ammonium
phosphate and (B) tetrahydrofuran. An A/B ratio of 9:1
was mixed and filtered through a 0.2um membrane.

It is noted that these analytical conditions were
not problem-free. Period column washing with water and
frequent pump dissembly and cleaning were necessary to
compensate for column and apparatus fouling that may
have been caused by higher molecular weight homologs in
the sodium lauryl sulfate additive.

The calibration standards included sodium form
polystyrene sulfonates obtained from Pressure Chemical
Co., Pittsburgh, Pa., and sodium toluene sulfonate.
Measurements were taken at 0.5 to 1.0ml/min flow rates.
The logarithm of the molecular weight of the standards
was linear it suggests a framework for approaching an
interpretion of the structure of the scission products.
This application of size exclusion chromatography
measurements must be viewed as a first approximation
because of the unmeasured differences between the
chromatographic behavior of the linear standards and
the expected branched structure of the scission
products.

RESULTS

Scission reactions were carried out with nominal
4, 8 and 12 mole %DVB where f = 0.04, 0.08 and 0.12,
respectively. The corresponding times required to
reach degelation were estimated as 4, 7.5 and 10 hours.
The time uncertainty of the degelation "point" is
estimated as 0.2 to 0.5 hr.

The treatment with hydrogen peroxide caused the
residual resin weight to decrease with time. The
weight of the 12 %DVB resin measured after drying was
observed to undergo a linear descent starting at 1.5g
and falling to zero at 10 hrs where degelation
occurred. The results are shown in Figure 1. This
shows that the intermediate scission pathway is a
macroscopically continuous process unmarked by abrupt
change in the chemical pathway. Fragmentation starts
at the beginning of the degradation and an accompanying
weight loss occurs until dissolution is complete.

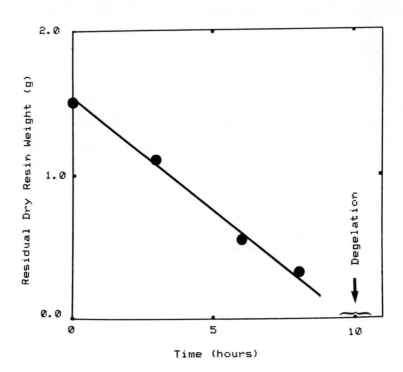

Figure 1. The measured dry weight after hydrogen peroxide
scission of sulfonated PSDVB (12% DVB cation exchange resin)
is seen to decrease linearly with reaction time. The time
obtained by extrapolating to zero weight corresponds to
visual observation degelation indicated by the disappearance
of the resin particles.

The aqueous size exclusion chromatograms obtained
for the three resins throughout their scission were
marked by an early appearance of a dominant peak at
molecular weight 200 Daltons. Similarly, all analyses
in the vicinity of the time of degelation were marked
by a major prominance with an estimated molecular
weight of 2000 Daltons or slightly larger. With X8 and
X12 we also observed a small peak of intermediate
molecular weight in the range between 400 and 900
Daltons.

The variation of the SEC analyses with time was
examined in detail with the X12 resin. The results are
shown in Figure 2. The early and sustained presence of
the molecular weight 200 peak indicates formation of
fragments whose molecular weight corresponds to the
pendant sulfonated aromatic rings (or a possibly
related degradation product) as an initially prominant
and subsequently continuing feature of the scission
process. Close inspection of the chromatograms showed
the appearance of an unresolved satellite peak of
variable apparent area corresponding to still smaller
size molecules.

Following the emergence of the preceding low
molecular weight peak, the degradation moved into
dominance by larger size fragments indicated by one
peak of 500-750 molecular weight accompanied by a
lesser peak with molecular weight in the range of
2000-2500. As the degradation moved into the half way
point and beyond, the relative amounts of material
represented by these two peaks were reversed, the
larger molecular weight being clearly dominant at the
time of degelation.

The molecular weight of these two peaks can be
compared to the reference values of MC = 873.2 and MX =
1987.6 calculated from Equations 2 and 1, respectively,
for f = 0.12.

The close correspondence between the fragments
molecular weights, 500-750 observed (873.2 calculated)
and 2000-2500 observed (1987.6 calculated), leads to
the conclusion that linear chain fragments and X-units
are apparently both formed after the scission process
begins.

The early formation of relatively small soluble
fragments of molecular weight 200 is followed by an
increasing amount of fragment of molecular weight 750
but the absence of the 2000-2500 molecular weights. At
the mid-point of the degradation the 2000-2500 molecular
weight peak arises and then dominates the degradation
product mixture.

The formation of small fragments with molecular
weight near 200 suggests that pendant group scission

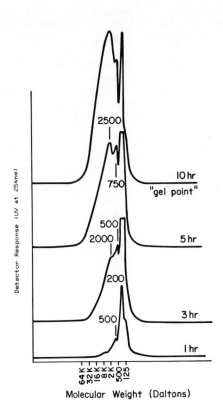

Molecular Weight (Daltons)

Figure 2. Size exclusion chromatograms of samples taken
during hydrogen peroxide degradation of the 12% DVB sample
whose mass depletion is shown in Figure 1. The potential
for formation of topogically significant scission fragments
is indicated. The apparent molecular weight at 200 Daltons
is close to that of the sulfonated pendant aromatic rings.
Peaks in the range 2000-2500 are near the calculated mass
(1978.6) of the average X-unit cell defined by Flory (6).
Peaks of approximate molecular weight 500-750 are comparable
to expected average intercross-link chain mass (873).

accompanies and probably precedes chain scission. Unit
scission is required to remove a pendant group
fragment. The delayed appearance of the 500-750 peak
and still later appearance of the 2000-2500 peak are
consistent with topological requirements that release
of a linear chain as a scission product requires two
cuts on the same chain while at least four cuts on four
contiguous chains are required to remove an X-unit.

It is worth noting that evidence is not apparent
for the formation of still larger molecular weight
fragments, referring to fragments whose molecular
weights would imply two or more X-units connected by a
single unbroken chain. A possible explanation for this
is that the formation of such larger fragments is
statistically less unlikely. Moreover, even if such
larger fragments were formed, it is possible that they
would be trapped within the network. Their eventual
release would be eclipsed by a diffusion impediment
that could enhance their remaining as stationary
targets for scission, for example, of a single
connecting chain that would form two sub-fragment
X-units. Once the latter cut were made, the possible
diffusional barrier would be lowered.

To summarize, the following hierarchy in the
formation of scission fragments is consistent with the
experimental results:

SCISSION FRAGMENT	NO. OF CUTS	ORDER OF APPEARANCE
pendant group	1	Initial
chains	2	Second
X-units (+)	4	Third
Higher (+-+, etc.)	6(etc)	Not observed

The present experiments may be subject to some
uncertainty in terms of molecular weight estimates and
diffusion effects that could affect the exactness of
these interpretations. The conclusion is reached that
the experiments strongly suggest evidence for
discontinuous topological quantification. In other
words, the order and rate of fragment release is
consistent with expectations based on fragment
topology.

Further study of network fragments will eventually require coping with the possibility of a larger range of scission fragments than have been identified here. The reason for expecting the added complexity stems from the fact that size exclusion chromatography is non-interactive and therefore has an obvious tendency to mask chemical differences between molecules of different composition but similar size. Even so, the potential for using the characterization of network fragments to probe the topological aspects of branched or crosslinked polymer structure emerges as an area that invites further study.

LITERATURE CITED

1.   Wood, W., J. Phys. Chem. 61, 832 (1957).

2.   Hookway, H.T., and Selton, B., J. Phys. Chem. 62, 493-4 (1958). ˇ

3.   Wiley, R., and Reich, E.,   J. Polym. Sci. A-1, 6, 3174-6 (1968).

4.   Dusek, K., and Prins, W., Adv. Polym. Sci. 6,1-102 (1969).

5.   Smith, W.B., and Temple, H.W., J. Phys. Chem. 72, 4613-9 (1968).

6.   Ziabicki, A., Polymer 20, 1373-1381 (1979).

7.   Flory, P.J., J. Phys. Chem. 11, 512 (1943).

RECEIVED October 31, 1983

Fractionation and Characterization of Commercial Cellulose Triacetate by Gel Permeation Chromatography

F. MAHMUD and E. CATTERALL[1]

Department of Applied Chemistry, Faculty of Applied Science, Coventry (Lanchester) Polytechnic, Coventry, England

Commercial cellulose triacetate samples were fractionated by both fractional precipitation and preparative gel permeation chromatography (GPC). The
triacetate fractions were characterized by visco-
metry, high speed membrane osmometry (HSMO) and
GPC. A fair agreement has been found between the
molecular weights of various triacetate fractions
determined by the three procedures.

All unfractionated cellulose triacetate samples
and high molecular weight fractions showed a shoulder on the high molecular weight side of the GPC
distribution. Material isolated from this region
was found to be highly enriched in mannose and
xylose, attributed to the presence of a hemicellulose derivative. Cellulose triacetate from cotton
linters did not show this behavior.

The universal calibration approach ([η].M vs
elution volume) for polystyrene standards and
narrow molecular triacetate fractions show slight
deviation from linearity. This departure from
linearity has been attributed to differences in
both hydrodynamic behavior and the Mark-Houwink
exponent 'a' for the two polymers in question.

A literature survey (1 - 11) on the fractionation of cellulose
triacetate by precipitation indicates that in most cases it has
been unsuccessful due to the possibility of hydrogen bonding between polymer and solvent in solutions. (10, 12). GPC has been
applied to the fractionation of cellulose derivatives by many
workers. Segal (13), Meyerhoff (14 - 16), Muller and Alexander
(17) have reported the fractionation of cellulose nitrate by
GPC. Muller and Alexander (17), Brewer, Tanghe, Bailly and Burr

[1]Current address: Petroleum and Gas Technology Division, Research
Institute, University of Petroleum and Minerals, Dhahran, Saudi
Arabia

0097–6156/84/0245–0365$06.00/0
© 1984 American Chemical Society

(18) have also used GPC for the fractionation of cellulose acetate and carbanilate respectively. Maley (19) and Cazes (20) reported some work on GPC fractionation of cellulose esters, but gave no data. It is worth mentioning here that the successful fractionation of cellulose triacetate has not been reported so far in the literature.

The prime object of the present study was to determine the compositional polydispersity of commercial cellulose triacetate and to examine the effect of molecular weight and molecular weight distribution on the mechanical properties of the fibres.

Experimental

Materials. Cellulose triacetate samples with 61.7 - 62% acetyl value, were all commercial grade and were supplied by Courtauld's Ltd., Coventry England. The chemicals and the solvents used in this work were all analytical grade materials.

Fractionation Procedures. 1. Fractional precipitation. A 10% (m/V) solution of a commercial grade triacetate sample was dissolved in 300 ml $\underline{N}$-methylpyrrolidone and 700 ml of acetone (30:70 V/V) and was thermostated for 2 hours at 25°C prior to the addition of 460 ml of petroleum ether (60-80°) as precipitant. The solution with the precipitant was gently warmed to 45°C to redissolve the precipitate and gradually cooled in the thermostat. Phase separation took place after a while, and the phases were isolated from each other by filtration. The gel like phase thus isolated was the first primary fraction. The subsequent fractions were isolated in the same way by the further successive additions of precipitant to the solution. The last fraction was isolated by the addition of a large volume of the precipitant and allowing the solution to stand for 72 hours before the phase separation is affected by filtration as stated above.

Seven primary cellulose triacetate fractions were isolated by this method. The first primary fraction rich in hemicellulose was redissolved and reprecipitated into three subfractions in the same way as described above. The refractionation of the first fraction was necessary to isolate the hemicellulose material for subsequent analysis and characterisation.

2. Gel Permeation Chromatography (GPC). Waters Associate Model 200 GPC was used with 4' x 3/8" 'styragel' columns with an internal diameter of 0.311" and refractometer detector. The basic characteristics and operation of the instrument have been previously described in detail (19-20). Some of the operating conditions used in this study are outlined below.

Column exclusion limits   : $7x10^5-5x10^6$, $5x10^6$, $5x10^3$ & $2-5x10^3$A

Mobile phase   : Dichloromethane

Flow rate   : 1 ml/min.

Sample concentration   : 0.5% m/V

Sample solution preparation: Allowed to stand overnight and then
filtered through glass sinter No. I
porosity.

Operating temperature   : Ambient

Injection volume   : 2 ml

Refractive index attenuator: X8(1/16" null glass)

Syphon size   : 5 ml

Choice of Solvent. N-Methylpyrrolidone (NMP) was initially used
as the mobile phase but proved to be unsatisfactory because of
(i) high solution viscosities, (ii) exceedingly small differences
in refractive index between NMP and cellulose triacetate solutions, (iii) erratic base line. In view of this dichloromethane
was employed. Some additional benefits derived from this mobile
phase are: (i) a decrease in elution volume due to low solution
viscosities, (ii) fast solvent recovery due to low boiling point
of dichloromethane and (iii) ease of obtaining preparative GPC
cuts of cellulose triacetate.

Preparative GPC of Cellulose Triacetate Sample. A 1% (m/V) solution of cellulose triacetate (medium) prefiltered through porosity 3 glass sinter was fractionated by repeated injection
through the column set described above. Seven cuts covering the
entire elution curve were collected. The flow rate, injection
time and the experimental conditions were identical to those
stated above.
   A total of 50 injections were made. Fractions were recovered
by removing dichloromethane under vacuum at low temperature. The
cuts were characterized in the same way as described previously
for cellulose triacetate fractions.

Calibration of Gel Permeation Chromatograph. The chromatographic
system was calibrated using:
(1) Polystyrene standards
(2) Narrow molecular weight cellulose triacetate fractions
(3) A 'universal' calibration approach

Polystyrene standards. Solutions of the monodisperse polystyrenes (Waters, Mass., USA) in N-methylpyrrolidone (0.5%m/V) and
dichloromethane (0.125% m/V) were used as calibrants. Figure 1
shows a plot of log (η) vs. log M_n for cellulose triacetate fractions in dichloromethane at 21 $^{\circ}$C.

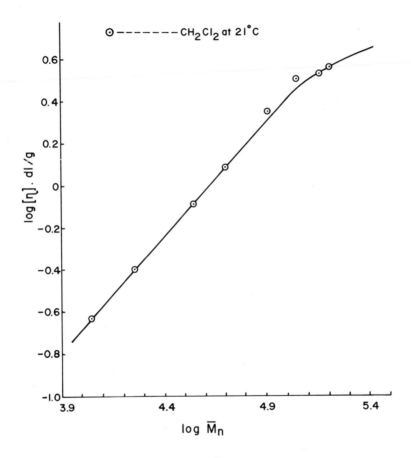

Figure 1. Log [η] versus log $\overline{M}_n$ relationship for cellulose
triacetate fractions.

Narrow Molecular Weight Triacetate Fractions. Narrow molecular
weight cellulose triacetate fractions were obtained by both fractional precipitation and preparative GPC as described above. The
number average molecular weight ($\overline{M}_n$) of the various fractions and
cuts was determined by high speed membrane osmometry. A linear
dependence of GPC elution volume on log molecular weight for all
cellulose triacetate fractions was found in both N-methylpyrrolidone and dichloromethane.

Universal Calibration. A function of the hydrodynamic volume $[\eta] \cdot M$
was plotted against the elution volumes of cellulose triacetate
fractions and polystyrene standards run in dichloromethane have
all indicated slight deviation from linearity as shown in Figure 2.

Discussion

Fractional Precipitation of Cellulose Triacetate. The reported
partial or non-fractionation of cellulose triacetate from chlorinated hydrocarbons or acetic acid may be explained in terms of the
polymer-solvent interaction parameter χ (1-11). The χ-values for
cellulose triacetate-tetrachloroethane and cellulose triacetate-
chloroform systems are reported (10,21) as 0.29 and 0.34 respectively. The lower values of χ for such systems will result in a
smaller or negative heat of mixing (ΔHm) and therefore partial or
non-fractionation of the polymer in question results.
     The poor fractionation from acetic acid has been attributed to
the intermolecular hydrogen bonding between solvent molecules and
thus a lesser polymer-solvent interaction. This means the total
heat evolved due to hydrogen bonding between polymer and solvent
molecules will be smaller than in the case of chloroform and tetra-
chloroethane and hence ΔHm (22) will be larger or more positive.
     The structural homogeneity of the various cellulose triacetate
fractions obtained by fractional precipitation was established by
both infrared and nuclear magnetic resonance spectroscopy.

Calibration of Gel Permeation Chromatograph Polystyrene Calibration.
A plot of molecular size in ($\mathring{A}$) versus elution volume for polystyrene standards in dichloromethane showed deviation from linearity
at about 2,200 $\mathring{A}$ which may be attributed to imperfect column resolution, peak broadening, axial dispersion and skewing. The extensive tailing of the chromatograms of high molecular weight polystyrene standards observed in dichloromethane has also been reported in the literature (23-26).

Narrow Molecular Weight Triacetate Calibration. A linear relationship was found when log $\overline{M}_n$ against the elution volumes of various
cellulose triacetate fractions was plotted. For narrow molecular
weight distribution triacetate fractions, the GPC experimental
average molecular weight, termed $\overline{M}_{peak}$ can be expected to conform
to the following equation

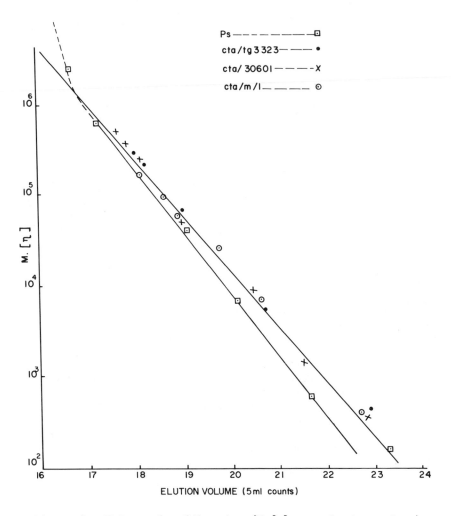

Figure 2.   Universal calibration (M·[η]   vs elution volume)
in dichloromethane at 21 C.

$$\bar{M}_{peak} \simeq \bar{M}_w \simeq \bar{M}_v \simeq \bar{M}_n$$

However, for unfractionated triacetate samples and for fractions of broader molecular weight distribution, this equation will not hold and therefore taking $\bar{M}_n$ or $\bar{M}_v$ as $\bar{M}_{peak}$ will lead to serious errors. This fact is evident from the results shown in Table I.

The apparent difference between viscosity average ($\bar{M}_v$) and number average molecular weight ($\bar{M}_n$) for unfractionated triacetate samples and high molecular weight fractions may be attributed to the presence of hemicellulose and polydispersity effect in these materials as shown in the table in question and in Figure 3.

Universal Calibration. Plots of $[\eta]\cdot M$ against elution volumes indicate that polystyrene and cellulose triacetate follow different calibrations as shown in Figure 2. This deviation from linearity may be due to the following reasons.

1. Linear polymers, polystyrene and cellulose triacetate exhibit differences in hydrodynamic behavior in solution. Cellulose and its derivatives are known to have highly extended and stiff chain molecules below a $\bar{D}p$ of about 300, but as the $\bar{D}p$ increases above 300 the chain tends to assume the character of a random coil (27,28). The assumption that hydrodynamic volume control fractionation in GPC may not be true for polystyrene and cellulose triacetate, though it has been found satisfactory for non-polar polymers in good solvents (29).

2. The Mark-Houwink exponent 'a' for cellulose triacetate in dichloromethane was found 1.10-1.14 compared to polystyrene with 'a' = 0.71. These values were obtained experimentally in the present work.

Parikh (12) has found higher values for the exponent 'a' using the following polymer-solvent systems:

   (i)   Cellulose triacetate-chloroform at $25^{\circ}C$: 'a' = 1.33
   (ii) Cellulose triacetate-tetrachloroethane at $25^{\circ}C$:'a'= 1.24
   (iii) Cellulose triacetate-Acetic acid at $25^{\circ}C$: 'a' = 1.18.

Though the 'a' values for cellulose triacetate-dichloromethane system appear high in the present study, it is still not surprising when compared to the above stated 'a' values reported by Parikh (12). The difference in the values of 'a' for polystyrene and cellulose triacetate may account partly for the deviation in slopes as shown in Figure 2.

Ethyl cellulose and cellulose triacetate have been shown to form hydrogen bonded associates with dichloromethane (10,30). If this is so, then cellulose triacetate-dichloromethane interaction will be favored over polystyrene-cellulose triacetate interaction and thus no adsorption should be expected.

The partial blocking of the GPC column with 5×10^5 Å exclusion limit in both dichloromethane and $\underline{N}$-methylpyrrolidone may be attributed to the presence of hemicellulose in both unfractionated triacetate samples and high molecular weight fractions used in this work. The blocking of the column in question was indicated

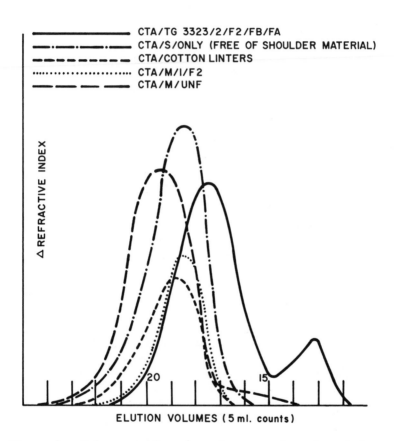

Figure 3.   Gel permeation chromatograms of various triace-
            tate samples and fractions.

Table 1. Analytical Data of Various Cellulose Triacetate Samples and Fractions

Sample	$21°C$ $[\eta]$ CH_2Cl_2	$\bar{M}_n$	$\bar{M}_v$	E.V. (5ml) (DCM)	$\bar{D}_p$	$\bar{M}_n$ (GPC) (DCM)	$\bar{M}_w$ (GPC) (DCM)	$\bar{M}_w/\bar{M}_n$ (DCM)
CTA/M/1/UNF	2.04	72000	105680	18.60	250	46000	121334	2.63
CTA/M/1/F1	2.88	114200	115880	18.22	396	97420	165639	1.70
CTA/M/1/F1/FA	3.20	228375	151360	17.80	793	1110474	217469	1.96
CTA/M/1/F1/FB	2.72	188720	137090	18.10	655	108525	186029	1.71
CTA/M/1/F1/FC	1.68	82210	83946	18.64	285	91907	109982	1.19
CTA/M/1/F2	1.95	111090	113760	18.22	385	91000	124957	1.37
CTA/M/1/F3	1.58	93916	85310	18.64	326	70362	99050	1.40
CTA/M/1/F4	1.38	87000	77804	19.00	302	60000	88982	1.48
CTA/M/1/F5	1.18	66300	67143	19.64	230	46620	55926	1.19
CTA/M/1/F6	0.60	35658	36658	20.60	123	19800	36158	1.80
CTA/M/1/F7	0.236	14170	16136	23.00	49	18600	19196	1.03

by a rapid rise in the system pressure which necessitated the removal of this column in order to overcome the problem stated above.

Meyerhoff (14-16) has also observed similar blocking of the gel column using cellulose trinitrate fractions with molecular weight above 1.4×10^6, while fractions with molecular weight 4.2×10^6 could not be separated. It is obvious from these results that he did not, however, realize the presence of the hemicellulose derivatives in the wood-pulp based cellulose nitrate and its role in blocking of the high porosity column as shown in this study.

Acknowledgments

Many people and organisation have contributed to this work, notably Courtauld's Ltd., Coventry, England.

Nomenclature of Cellulose Triacetate Samples and Fractions

Samples: In CTA/S/UNF, CTA/M/UNF, CTA/3060/UNF and CTA/TG 3323/UNF CTA stands for cellulose triacetate. The designation of S,M,3060 and TG 3323 are batch numbers given to these samples by Courtauld's Ltd. UNF is the symbol for unfractionated sample.

Fractions: Each triacetate fraction, like the respective sample, starts with the symbol CTA (first column from left to right) and is then followed by the batch number (2nd column), fractionation number (3rd column), fraction number (4th column) and sub-fraction number (5th column respectively).

Literature Cited

1. Elod, E. and Schmidt-Bielenberg, J. Physik. Chem. B25, 38, 1934.
2. Lachs, H.J., Kolloid, J., 79, 91, 1937.
3. Levi, G. R., Gazz. Chim. Ital, 68, 589, 1938.
4. Bezzi, S., Atti 1st. Veneto Sci., 99, 905, 1939-40; Chemical Abstracts, 38, 1356, 1944.
5. Munster, A., J. Polym. Sci., 5, 58, 1950.
6. Cumberbirch, R. J. E., Shirley Institute Memoirs, 31, 1958.
7. Thinius, K., Plaste Kautschuk, 6, 547, 1959.
8. Okunev, P.P. and Tarakanov, O. G., Vysokomol, Soed., 4, 5, 688, 1962.
9. Dymarchuk, N.P., Zhurnal Prikladnoi Khimi, 37, No. 10, pp 2263-2268, English Translation, October 1964.
10. Howard, P., and Parikh, R. S., J. Polym. Sci., A-1, 4, 407-418, 1966.
11. Geller, B. E., Khimicheski Volokna, 11, No. 5, pp 1-6, English Translation, 1969.
12. Parikh, R.S., Ph.D. Thesis, University of Surrey, 1965.
13. Segal, L., J. Polym. Sci., B4, 1011, 1966.
14. Meyerhoff, G., Makromolek. Chem., 89, 282, 1965.
15. Meyerhoff, G. and Jovanovic, S., J. Polym. Sci., B5, 495, 1967.
16. Meyerhoff, G., Makromolek. Chem., 134, 129, 1970.
17. Muller, T. E., and Alexnader W. J., in"Analytical Gel Permeation Chromatography"(J. Polym. Sci. C, 21), Johnson, J. F. and Porter, R.S., Eds., Interscience, New York, p. 283, 1968.
18. Brewer, R. J., Tanghe, L. J., Bailey, S. and Burr, J. T., J. Polym. Sci., A-1, 6, 1697, 1968.
19. Maley, L. E., Analysis and Fractionation of Polymers, J. Polym. Sci., C-8, 253-268, 1965.
20. Cazes, J., J. Chem. Educ., 43, A567, 1966.
21. Hager, O. and Vander Wyk, A. J. A., Helv. Chim. Acta, 1940 23, 484.
22. Hildebrand, J. and Scott, R., "The Solubility of Non-Electrolytes," 3rd Ed., Reinhold, 1949.
23. Harmon, D. J., in "Analysis and Fractionation of Polymers," J. Polym. Sci., C-8, Mitchell, J., Jr. and Billmeyer, F.W., Jr., Eds., Interscience, New York, p. 243, 1965.
24. Tung, L. J., J. Polym. Sci. 10, 375, 1261, 1274, 1966.
25. Hess, M. and Kratz, R. F., J. Polym. Sci. A-2, 4, 73, 1966.
26. Smith, W. N., J. Appl. Polym. Sci., 11, 639, 1967.

27. Flory, P. J., "Principles of Polymer Chemistry," Cornell
     Univ. Press, Ithaca, New York, Chap. 13, 1953.
28. Stamm, A. J., "Wood and Cellulose Science," Ronald, New York,
     pp 96, 108, 1966.
29. Dawkins, J. V., Br. Polym. J., 4, 87–101, 1972.
30. Brookshaw, A. P., Br. Polym. J., 5, 229–239, 1973.

RECEIVED December 12, 1983

INDEXES

Author Index

Subject Index

Production by Paula Bérard
Indexing by Florence Edwards
Jacket design by Anne G. Bigler
based on an idea by Ann Kah

Elements typeset by Hot Type Ltd., Washington, D.C.
Printed and bound by Maple Press Co., York, Pa.

RETURN TO: CHEMISTRY LIBRARY

100 Hildebrand Hall • 510-642-3753

LOAN PERIOD 1	2 *1 Month*	3
4	5	6

ALL BOOKS MAY BE RECALLED AFTER 7 DAYS.

Renewals may be requested by phone ~~or, using GLADIS, type inv followed by your patron ID number~~.

DUE AS STAMPED BELOW.

FORM NO. DD 10
3M 7-08

UNIVERSITY OF CALIFORNIA, BERKELEY
Berkeley, California 94720–6000